SpringerBriefs in Molecular Science

Chemistry of Foods

Series editor

Salvatore Parisi, Palermo, Italy

More information about this series at http://www.springer.com/series/11853

Daniele Pisanello

Chemistry of Foods: EU Legal and Regulatory Approaches

 Springer

Daniele Pisanello
Lex Alimentaria Studio Legale
Taviano (Lecce)
Italy

ISSN 2199-689X   ISSN 2199-7209   (electronic)
ISBN 978-3-319-03433-1   ISBN 978-3-319-03434-8   (eBook)
DOI 10.1007/978-3-319-03434-8

Library of Congress Control Number: 2014948750

Springer Cham Heidelberg New York Dordrecht London

Printed on acid-free paper

Springer is part of Springer Science+Business Media (www.springer.com)

Contents

Abbreviations

3-MCPD	3-monochloropropane-1,2-diol
ADI	Acceptable daily intake
ANS	Panel on food additives and nutrient sources added to food
Cd	Cadmium
CEC	Council of the European Communities
CEF	Panel on Food Contact Materials, Enzymes, Flavourings and Processing Aids
CEN	European Committee for Standardization
CONTAM	Panel on Contaminants in the Food Chain
CVMP	Committee for Medicinal Products for Veterinary Use
DMF	Dimethylfumarate
DON	Deoxynivalenol
DU	Downstream user
E 104	Quinoline yellow
ECHA	European Chemicals Agency
EEC	Economic European Community
EFSA	European Food Safety Authority
EREN	Emerging Risks Exchange Network
EU	European Union
FBO	Food Business Operator
FPM	Food Packaging Material
FUMO	Fumonisin
GAP	Good Agricultural Practice
GFL	General Food Law Regulation (EC) No 178/2002
GMO	Genetically Modified Organism
GMP	Good Manufacturing Practice
GPSD	Council Directive 92/59/EEC
Hg	Mercury
Hg^{2+}	Mercuric Mercury
ML	Maximum Level

MRL	Maximum Residue Limit
OTA	Ochratoxin A
PAH	Polycyclic aromatic hydrocarbon
Pb	Lead
PCB	Polychlorinated biphenyl
PCDD	Polychlorinated dibenzo-para-dioxin
PCDF	Polychlorinated dibenzofuran
PMTDI	Provisional Maximum Tolerable Daily Intake
PTWI	Provisional Tolerable Weekly Intake
RAPEX	Rapid alert system for non-food dangerous products
RASFF	Rapid alert system for food and feed
REACH	Registration, Evaluation, Authorisation and Restriction of Chemicals
SCF	Scientific Committee on Food
SCFCAH	Standing Committee on the Food Chain and Animal Health
Sn	Tin
SVHC	Substance of very high concern
TDI	Tolerable daily intake
TEF	Toxic equivalent factor
TFEU	Treaty on the Functioning of the European Union
TWI	Tolerable weekly intake
ZON	Zearalenone

Chapter 1
Food Safety in Europe. Law Bases

Abstract The single market of the European Union assures that the internal trade in goods, services, people, and money among the Member States are free from protectionist actions and other limiting measures. After the Maastricht Treaty, the European Union aims also at protecting the public safety, the animal and plant health, the environment and consumer's protection. For these reasons, the assurance of high food safety perspectives and the creation of free and transparent trade relations in the European Union can be obtained by means of a peculiar 'comitology' system. This Chapter aims to explain the role of comitology in the framework Regulation (EC) No 178/2002 and in all food safety issues and related regulatory documents.

Keywords Comitology · EFSA · European union · Examination procedure · Food business operator · Food safety · General food law · RASFF

1.1 Food Safety in Europe. An Introduction

The European Union (EU) single market assures that the internal trade in goods, services, people, and money among the Member States are free from protectionist actions and other limiting measures [14].

Over the last decades, a substantial corpus of European legislation has been adopted in order to secure the free movement of goods in the single market. Since the adoption of the Council Resolution of 1985 [5], the EU is engaged on the prevention of new barriers to trade, mutual recognition, essential requirements in the harmonised legislation, european standards developed by standardisation bodies and the notification of draft technical regulations. A significant part of this legislation concerns safety aspects of food products.

After the Maastricht Treaty (formally, the Treaty on European Union, signed on 7 February 1992), the EU policy aims also at protecting the public safety, the animal and plant health, the environment and consumer's protection. The first Directive on general product safety, the Council Directive 92/59/EEC of 29 June

© The Author(s) 2014
D. Pisanello, *Chemistry of Foods: EU Legal and Regulatory Approaches*,
SpringerBriefs in Chemistry of Foods, DOI 10.1007/978-3-319-03434-8_1

1992 (hereinafter, GPSD) was adopted in order to set mandatory requirements for protecting consumers' health and safety and to ensure the proper functioning of the internal market. The GPSD, now repealed by Directive 2001/95/EC [9], is intended to ensure a high level of product safety throughout the EU for consumer products that are not covered by specific sector norms (e.g. toys, chemicals, cosmetics, machinery). It also complements provisions of sectoral legislation which do not cover certain matters, for instance in relation to producers' obligations and authorities' powers and tasks.

The above mentioned Directive provides a generic definition of 'safe product'. Food products must comply with this definition: the free circulation within the single EU market is guaranteed when the safety of foods can be demonstrated. Accordingly to the GPSD, 'safe product' is defined 'any product which, under normal or reasonably foreseeable conditions of use, including duration, does not present any risk or only the minimum risks compatible with the product's use, considered as acceptable and consistent with a high level of protection for the safety and health of persons, taking into account the following points in particular [3]:

- The characteristics of the product, including its composition, packaging, instructions for assembly and maintenance
- The effect on other products, where it is reasonably foreseeable that it will be used with other products
- The presentation of the product, the labelling, any instructions for its use and disposal and any other indication or information provided by the producer, the categories of consumers at serious risk when using the product, in particular children'.

At present, 'risk' and 'category of risk' are not legally defined terms according to the GPSD. However, only risks and categories of risk for human health and safety are relevant in this ambit according to Article 2(b) of the GPSD. Examples of types of risk that are covered include chemical, mechanical, thermal and electrical risks, noise and flammability. On the other hand, environmental risks, dangers for animal and plant health and financial problems, are not included.

The provisions of this Directive are intended as a general legal framework to apply in so far as there are no specific provisions in European laws governing the safety of the products concerned. Thus, as long as EU provides for new regulatory framework dealing with emerging product-related risks or horizontal matters—chemical risk as regulated by the Regulation on Registration, Evaluation, Authorisation and Restriction of Chemicals (REACH) adopted on 2006—the original scope of GPSD will be progressively reduced.

Under GPSD ruling, the safety of a product must be assessed in accordance with: (a) European standards; (b) Community technical specifications; (c) codes of good practice; (d) the state of the art and consumers' expectations.

The GPSD also provides for an alert system—the Rapid Alert System for non-food dangerous products (RAPEX) system—between Member States and the European Commission. The RAPEX system ensures that the relevant Authorities are rapidly informed of dangerous products. Subject to certain conditions, Rapid

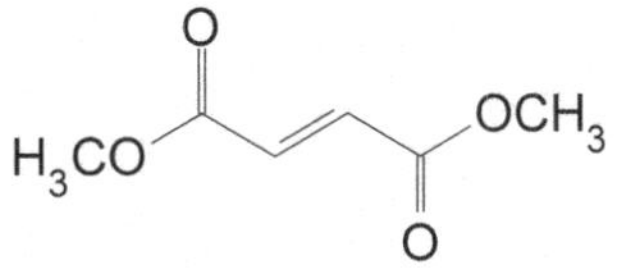

Fig. 1.1 The chemical structure of dimethylfumarate (*DMF*), molecular formula: $C_6H_8O_4$, molecular weight: 144.127 g/mol. DMF is a known biocide with recognised action against moulds. On the other hand, this chemical is also known for adverse skin contact reactions with following consequences: dermatitis, including itching, irritation, redness and burns. BKchem version 0.13.0, 2009 (http://bkchem.zirael.org/index.html) has been used for drawing this structure

Alert notifications can also be exchanged with non-EU countries. In the case of serious product risks, the Directive provides for temporary Decisions to be taken on Community-wide measures.

Under certain circumstances, the Commission may adopt a formal Decision requiring the Member States to ban the marketing of a product deemed posing a serious risk. With reference to suspected products, specific measures include the recall from consumers or the withdraw from the market [13]. Such decisions can be taken at Community level:

- Where Member States may be allowed to manage food-related risks in different ways
- Where a specific food safety or hygiene risk has to be managed immediately and EU laws cannot be urgently applied or are inadequate
- Where a specific and serious food-related risk can be eradicated by means of an EU measure.

At present, four decisions of this kind have been taken at Community level. One of these measures was related to temporary ban toys and childcare articles because of their content of phthalates which are substances used to soften plastics [3, 19]. The seriousness of safety concerns was later backed by a number of risk assessments which gave ground to a permanent ban now included in the Annex XVII of REACH Regulation [11], entries No 51 and 52.

Later, on 17 March 2009, the Commission adopted the Decision 2009/251/EC with relation to the general EU ban on all consumer products containing dimethylfumarate (DMF).

DMF (Fig. 1.1) is a biocide preventing moulds [22]: it may deteriorate leather furniture or footwear during storage or transport when relative humidity is notable. DMF has been often contained in little pouches: these articles can be fixed inside the furniture or added to the footwear boxes. As a result, DMF may evaporate and impregnate the product, protecting it from moulds. However, the action of DMF against consumers who can be in contact with products is known [21]. DMF can penetrate through the clothes onto consumers' skin with consequent skin contact dermatitis, including itching, irritation, redness and burns [21]. Acute respiratory troubles have been also reported in the literature [23]. The dermatitis may be particularly difficult to treat. Because of the seriousness of correlated health risks, a restriction dossier prepared under REACH has confirmed the need for a permanent ban of DMF. This substance has been included on 15 May 2012 in the Annex XVII of REACH Regulation, including a risk assessment and a socio-economic analysis.

In addition to the basic obligation to place only safe products on the market, food producers must inform consumers of every risk associated with supplied products. They must take appropriate measures to prevent such risks and be able to trace dangerous products.

1.2 Setting EU Mandatory Requirements for the Single Market. An Overview on the Role of Comitology

The creation and the implementation of safety standards and requirements at the EU level can be defined as the result of a process with the aim of assuring high safety perspectives, a reliable protection for consumers, free and transparent trade relations. In general, the main role is played by the services of the EU Commission. In fact, the Commission holds the exclusive right to draft legislative proposals and is thus a co-legislator. It is responsible for the implementation of EU policies and it is also the main actor in the executive branch of the EU. It supervises that EU treaties and related legislation are respected by Member States. Moreover, the Commission negotiates on behalf of the EU in many international settings as an external representative.

Broadly speaking, the EU Commission must consult a committee before the implementation of an EU legal act: every EU country has to be represented [12]. The committee is requested to furnish an opinion without compulsory effects on the final decision of the Commission.

The above mentioned system, namely defined 'comitology', may be roughly traced in the early '60s: in fact, the first example is commonly considered in connection with the implementation of common EU agricultural policies. This strategy can be considered as a convenient compromise between National authorities and the EU Commission with relation to the exercise of political powers. In general, the Commission is requested by the Council of Ministers to consult committees of Member States representatives when adopting secondary rules in several policy areas.

In detail, the 'comitology' system—codified by the Single European Act in 1987 and ruled by Regulation (EU) No 182/2011—is based on the creation of single forums for discussion, also named 'committees' [12]. Every committee is composed of several representatives from Member States. Anyway, all committees are chaired by the EU Commission. In summary, the main role of these organisms is to establish the dialogue between the EU Commission and National administrations before adopting implementing measures [12]. In detail, working rules of committees are mentioned in the Council Decision 1999/468/EC, repealed by Regulation (EU) No 182/2011 [12].

With relation to basic aims of this book, it should be remembered that the 'Treaty of Lisbon' has recently modified the framework for implementing powers that are conferred upon the Commission by the legislator. This document [15] has been written an approved with the aim of amending previous Treaties on European Union and the European Community. In detail, the new Treaty has recognised the fundamental difference between the power of adopting non-legislative acts of general application or amending non-essential elements of legislative acts (delegated acts) and the formal adoption of implementing acts.

With reference to the last power, Member States can control the Commission's exercise of implementing powers in accordance with Article 291 of the Treaty on the Functioning of the European Union, while this legal ability is not ascribed to the European Parliament and the Council. On the other hand, the legislator can control the exercise of the Commission's powers on delegated acts by means of revocation and/or objection rights, in accordance with Article 290 of the Treaty on the Functioning of the European Union.

On these bases, the adoption of every implementing act by means of the intermediate 'draft act' can be made on condition that certain criteria are applied and respected. One of these conditions concerns specifically the referral to a dedicated 'appeal committee' with relation to the necessity of giving opinions before the approval of the final draft act by the Commission.

This explanation may be considered necessary because the comitology system requires that a specific process—the 'examination procedure'—has to be applied in accordance with Article 2.2 of the GPSD [12]:

- A legal act identifies the need for uniform conditions of implementation, and
- The nature or the impact of the implementing act concerns also 'the common agricultural and common fisheries policies', 'the environment, security and safety, or protection of the health or safety, of humans, animals or plants', and 'the common commercial policy'.

Actually, another strategy—the so-called 'advisory procedure'—may be applied for the adoption of implementing acts in duly justified cases.

Anyway, the examination procedure can be carried out if the Commission is assisted by a committee composed of representatives of Member States (the same provision is usual also for common advisory procedures). The committee is chaired by a representative of the Commission without roles in the committee except for leading work within the committee. Should this committee express a positive opinion, the Commission would adopt the draft implementing act. Should this committee express negative opinions, the adoption would be blocked, while the absence of opinions would give the Commission the right of adopting the draft except for some cases including also acts with relation to the protection of the health or the safety of humans, animals or plants, or definitive multilateral safeguard measures. In addition, the referral of an appeal committee is guaranteed for subsequent amendments.

1.3 Setting EU Mandatory Requirements for the Single Market. Comitology and Food Policy

With specific regards to foods, the framework Regulation (EC) No 178/2002 (hereinafter, General Food Law, or GFL) provides for the establishment of the 'Standing Committee on the food chain and animal health' (SCFCAH). This committee has a mandate covering the entire food supply chain [10], 'from farm to fork'. SCFCAH has different competencies, ranging from veterinary issues to foodstuffs and feedingstuffs. Before the GFL, these competencies were originally ascribed to the Standing Veterinary Committee,[1] the Standing Committee for Foodstuffs[2] and the Standing Committee for Feedingstuffs.[3]

In addition, SCFCAH can give opinions about plant protection products and the setting of maximum residue levels. In detail, SCFCAH is organised in specific sections in accordance with the Article 58 of GFL. Every section is named with the correlated argument or topic; consequently, SCFCAH is subdivided in the following sections [10]:

- General food law
- Biological safety of the food chain
- Toxicological safety of the food chain
- Controls and import conditions
- Animal nutrition
- Genetically modified food and feed and environmental risk
- Animal health and animal welfare
- Phytopharmaceuticals.

Moreover, SCFCAH is entitled to examine the application of traceability (GFL, Article 17) and implementing measures for rapid alerts (Sect. 1.5) in accordance with Article 5 of the Regulation No 182/2011. Finally, emergency measures for food and feed of Community origin or imported from a third country are also ascribed to SCFCAH (GFL, Article 53).

The political importance of SCFCAH is mainly correlated to the necessity of setting a sort of mediation organism when a potential breach to GFL or possible barriers to internal market arise from a measure taken by another Member State in the field of food safety. Should one or more of these situations occur, SCFCAH would have the right of forwarding the matter to the Commission with the subsequent advice to the interested Member State, until an agreement can be reached or a specific opinion on relevant contentious scientific issues may be requested (GFL, Article 60).

---

[1] Established by Council Decision of 15 October 1968 setting up a Standing Veterinary Committee (68/361/EEC), now repealed by Reg. (EC) No 178/2002.

[2] Established by Council Decision of 13 November 1969 setting up a Standing Committee for Foodstuffs (69/414/EEC), now repealed by Reg. (EC) No 178/2002.

[3] Established by of 20 July 1970 setting up a Standing Committee for Feeding-stuffs (70/372/EEC), now repealed by Reg. (EC) No 178/2002.

Besides SCFCAH, the GFL provides also for an advisory committee which co-operates with the Commission in order to assure the open and transparent public consultation, directly or through representative bodies, during the preparation, the evaluation and the revision of food law, except where the urgency of the matter does not allow it [10].

Anyway, all issues of food legislation in the EU consultation system concern different aspects relating to the nutritional labelling and the presentation of food and feed products, food and feed safety, the human nutrition, the animal health, plant protection and conditions for the marketing of seed and propagation material, including biodiversity and the industrial property.

1.4 Setting National Requirements Within the EU. The Notification Directive

One of the most difficult and important tasks in the management of the internal market is correlated to the necessity of avoiding the adoption of national technical regulations and standards which may create new barriers to trade activities. One of fundamental tools in preventing the fragmentation of the EU single market is the Directive 83/189/EEC, also named 'Notification Directive' [4], codified by Directive 98/34/EC [8].

At present, the 'Notification Directive' covers also all agricultural and industrially manufactured products and possible notifications about new technical norms or regulations with national importance. In addition, the Directive 98/34/EC is now applied in all EU Countries and in Turkey (partially) because of the Association Agreement between this Nation and the EU [6, 7].

In summary, the 'Notification Directive' concerns:

- The procedure for the provision of information on new national norms, standards and similar documents. For these topics, following organizations are involved: national standardisation bodies, European Committee for Standardisation, European Telecommunications Standards Institute, European Committee for Electrotechnical Standardisation and the EU Commission. Actually, the procedure is carried out by European standardisation bodies by means of an annual contract
- The correct and transparent exchange of information in the regulatory field, with the aim of avoiding preventively that new possible technical regulations are adopted by Member States in spite of the necessary harmonization and without adequate 'standstill periods' (the suspension of the national draft legislation is required for a certain period to facilitate the discussion at Community level).

With reference to food products, all rules concerning food product composition, labelling, name, testing, life cycle through to disposal, etc., concern the application of the Notification Directive.

1.5 EU Food Law. General Remarks

The GPSD concerns all the EU product safety legislation [9] with relation to consumer products that are not entirely ruled by sectoral Directives (examples: toys, lighters). It has to be remembered that the EU food-related legislation is sectoral and is recognised to have priority over general provisions because of the main objective: the 'free trade of safe foodstuffs'.

As above mentioned, the framework of EU feed and food law is the GFL [10]. In accordance with this document, food business operators (FBO) are responsible for the safety and the compliance of foods with relation to relevant food legislations such as the Regulation (EC) No 882/2004 on official controls performed to ensure the verification of compliance with feed and food law, animal health and animal welfare rules. This responsibility is extended right along the whole food chain.

The adoption of the GFL has determined numerous innovations. Briefly, main pilasters of the new legislation can be summarised as follows [10]:

• In accordance to the Article 3, the food law cover 'food in general, and food safety in particular, whether at Community or national level; it covers any stage of production, processing and distribution of food, and also of feed produced for, or fed to, food-producing animals'
• In accordance with Article 5(1), the attention is mainly focussed on following elements: 'safety management, protection of consumers' interests, fair practices and, where appropriate, the protection of animal health and welfare, plant health and the environment'.

In addition, the approach to 'risk analysis' is specifically mentioned (GFL, Articles 6–10) with other principles: the so-called 'precautionary principle', the protection of consumers' interests and the importance of transparency [16]. This statement has to be made operational by means of independent scientific and technical evaluations. For this reason the GFL has established an independent Authority dealing with risk assessment: the European Food Safety Authority, hereinafter EFSA, (GFL, Articles 22–49).

The GFL introduces also new responsibilities and obligations for FBO (Articles 17–21): general safety requirements, product presentation, traceability, and crisis management tools [10].

Finally, the establishment of a specific rapid alert system for foods, named 'Rapid Alert System for food and feed' (RASFF), has to be considered because of the huge impact on the food safety in the EU. This system has really changed the perception of food safety actions in the EU because of the rapid exchange of information (GFL, Articles 50–52) on risks posed by food or feed and correlated measures [18].

1.6 EU Food Law. Connections Between Food Safety and Chemical Additives

As above mentioned, general rules of the GFL have to be integrated with a number of sectoral legislations laying down specific requirements relating to certain products. Several examples are: baby foods, supplements, products of animal origin, product of plant origin, etc. In addition, sectoral legislations are necessary when speaking of certain issues: food hygiene, food contact materials, contaminants, etc.

In other terms, the basic aim it to produce 'safe foods'. On the other hand, the concept of 'unsafe food' is not legally defined in EU laws since the adoption of GFL. Actually, the definition of 'unsafe product' has been provided by GPSD; moreover, a number of sectoral legislations dealing with specific health issues may be found. Once more, a general definition applying to unsafe foods as a whole does not appear to exist.

Basically, foods may be defined by GFL as edible substances or products for human consumption. The distinction between processed, partially processed or unprocessed foods does not appear important at this stage. On the opposite hand, the EU legislator considers one key issue[4]: unsafe foods cannot be placed

---

[4] Article 14 GFL reads as follows: 1. *Food shall not be placed on the market if it is unsafe.* 2. *Food shall be deemed to be unsafe if it is considered to be:* (a) *injurious to health;* (b) *unfit for human consumption.* 3. *In determining whether any food is unsafe, regard shall be had:* (a) *to the normal conditions of use of the food by the consumer and at each stage of production, processing and distribution, and* (b) *to the information provided to the consumer, including information on the label, or other information generally available to the consumer concerning the avoidance of specific adverse health effects from a particular food or category of foods.* 4. *In determining whether any food is injurious to health, regard shall be had:* (a) *not only to the probable immediate and/or short-term and/or longterm effects of that food on the health of a person consuming it, but also on subsequent generations;* (b) *to the probable cumulative toxic effects;* (c) *to the particular health sensitivities of a specific category of consumers where the food is intended for that category of consumers.* 5. *In determining whether any food is unfit for human consumption, regard shall be had to whether the food is unacceptable for human consumption according to its intended use, for reasons of contamination, whether by extraneous matter or otherwise, or through putrefaction, deterioration or decay.* 6. *Where any food which is unsafe is part of a batch, lot or consignment of food of the same class or description, it shall be presumed that all the food in that batch, lot or consignment is also unsafe, unless following a detailed assessment there is no evidence that the rest of the batch, lot or consignment is unsafe.* 7. *Food that complies with specific Community provisions governing food safety shall be deemed to be safe insofar as the aspects covered by the specific Community provisions are concerned.* 8. *Conformity of a food with specific provisions applicable to that food shall not bar the competent authorities from taking appropriate measures to impose restrictions on it being placed on the market or to require its withdrawal from the market where there are reasons to suspect that, despite such conformity, the food is unsafe.* 9. *Where there are no specific Community provisions, food shall be deemed to be safe when it conforms to the specific provisions of national food law of the Member State in whose territory the food is marketed, such provisions being drawn up and applied without prejudice to the Treaty, in particular Articles 28 and 30 thereof.*

on the EU market (GFL, Article 14.1) with the exclusion of primary productions for private domestic use or the use of food for private domestic consumption (GFL, Article 1.3).

The concept of 'unsafety' or injuriousness relates to the potential to harm human health. This concept is quite broader than that of 'defective product' as the latter is based on a direct negative impact on the health of the consumer (poisoned or contaminated food for example). On the contrary, injuriousness is broader as it may include toxic effects and health threats on coming generations: the contamination by dioxins can be a useful example. Another important matter concerns the possibility of allergenic reactions for specific categories of consumers [17].

Anyway, every identified hazard with possible concerns for the human health has to be managed by means of a correct and reliable risk assessment (GFL, Articles 14.3 and 14.4).

With the exclusion of microbiological hazards (example: *Salmonella* spreading) and physical risks (example: presence of glass fragments into foods and beverages), the category of chemical hazards has to be taken into account by FBO. This 'category' includes:

- Possible allergenic reactions [17]
- Possible detection of hazardous chemicals into foods.

The problem of the food intolerance should be discussed in a broadest ambit of food science and hygiene by different viewpoints. With relation to the detection of 'hazardous chemicals', it should be also considered that:

- These substances may be naturally present in the original food or ingredient
- These molecules may be originated by microbial spreading or incipient contamination
- The detection of 'unsafe' substances may depend on the external contamination by packaging materials or by addition of food-grade ingredients with some impurity
- Anyway, the risk may be correlated with a peculiar and legal limit depending on the peculiar historical moment, the available knowledge, best available detection technologies and the specificity of national or international laws.

For these reasons, the establishment of an independent and transparent Authority, the EFSA, has been made necessary in recent times.

1.7 EFSA and Chemicals

EFSA has been established by means of the GFL with the aim of providing scientific opinions and technical supports for the EU legislation and correlated policies with specific reference to food and feed safety. At present, two different EFSA organisms are committed to perform risk management and risk assessment in a separate way: the Management Board and the Advisory Forum [2]. With reference to the Advisory Forum, this organism gives independent and reliable scientific opinions about risk assessment in different areas. Because of the variety of possible topics, the Advisory Forum is subdivided in Panels: each Panel is composed of expert consultants and representatives of regulatory bodies of Member States. Three of these divisions and units play a major role in the risk assessment with concern to chemicals in foods.

Firstly, the EFSA Panel on Contaminants in the Food Chain (CONTAM) is entitled to perform risk assessment-oriented evaluations on contaminants in food and feed. In detail, CONTAM has to assess whether the correlation between adverse health effects on the human being and the exposure to a specific chemical contaminant in foods can be reliable and demonstrable in the EU. The same thing has to be evaluated when speaking of adverse health effects in farm animals, fish and pets. Should this correlation be demonstrable, the subsequent step would be the examination of the worst scenario: the occurrence of safety risks to the consumer of foods of animal origin.

For these reasons, EFSA often launches public calls for data on the occurrence of contaminants in food and feedstuffs. At the same time, Member States and other interested stakeholders are invited to submit data. Should the evaluation be workable, the CONTAM Panel would have the possibility of establishing a health-based guidance value such as tolerable daily intakes for a substance such as genotoxic and carcinogenic chemicals. Should collected data be inadequate, the same CONTAM Panel would opt for the margin of exposure approach [20]. Between 2003 and 2013, the CONTAM Panel has carried out more than 100 scientific outputs [1].

With relation to regulated food ingredients and packaging materials and objects, two separated scientific Panels are entitled to make their own separated risk evaluations with the support of the 'Food Ingredients and Packaging' Unit:

- The Panel on Food Additives and Nutrient Sources Added to Food (ANS) can discuss and evaluate safety questions with relation to the addition of food additives, nutrient sources and other substances to food, except for flavourings and enzymes
- The Panel on Food Contact Materials, Enzymes, Flavourings and Processing Aids (CEF) can discuss every possible question concerning the safety of food packaging materials, objects and components. In addition, this Panel can discuss safety implications of enzymes, flavourings, processing aids and processes when correlated to the production of foods.

It has to be highlighted that the Panels' risk assessment work is correlated to the context of authorisation procedures for chemicals: in fact, only regulated substances may be added or used for food production and/or packaging in the EU. This approach obliges food players to be aware of pre-existing EFSA evaluations and the consequent authorization in the EU.

Finally, the role of the Dietary and Chemical Monitoring Unit should be remembered with relation to the risk assessment in the area of chemicals in foods. This EFSA unit collects, collates and makes a deep analysis of known data about food composition and consumption. In addition, this unit is entitled to make risk assessment-oriented analyses about food exposure in the EU. The basic aim of this activity is to provide reliable estimations of the human dietary exposure to a hazard through combining occurrence levels in a food with related consumption levels.

References

1. Alexander J, Benford D, Boobis A, Eskola M, Fink-Gremmels J, Fürst P, Heppner C, Schlatter J, van Leeuwen R (2012) Risk assessment of contaminants in food and feed. EFSA J 10(10):s1004. doi:10.2903/j.efsa.2012.s1004
2. Benford D, Bolger PM, Carthew P, Coulet M, DiNovi M, Leblanc JC, Renwick AG, Setzer W, Schlatter J, Smith B, Slob W, Williams G, Wildemann T (2010) Application of the margin of exposure (MOE) approach to substances in food that are genotoxic and carcinogenic. Food Chem Toxicol 48:S2–S24. doi:10.1016/j.fct.2009.11.003
3. Commission of the European Union (1998) Decision 1999/815/EC concerning measures prohibiting the placing on the market of toys and childcare articles intended to be placed in the mouth by children under three years of age made of soft PVC containing certain phthalates. Off J Eur Union L 315:46–49
4. Council of the European Communities (1983) Council directive 83/189/EEC of 28 March 1983 laying down a procedure for the provision of information in the field of technical standards and regulations. Off J Eur Comm L 109:8–12
5. Council of the European Union (1985) Council Resolution 85/C 136/01 of 7 May 1985 on a new approach to technical harmonization and standards. Off J Eur Comm C136:1–9
6. EC-Turkey Association Council (1996) Decision 1/95 of the EC-Turkey Association Council of 22 December 1995 on implementing the final phase of the customs union. Off J Eur Comm L35:1–46
7. European Economic Community and Turkey (1973) Agreement establishing an Association between the European Economic Community and Turkey. Off J Eur Comm C 113:1–77
8. European Parliament and Council (1998) Directive 98/34/EC of the European Parliament and of the Council of 22 June 1998 laying down a procedure for the provision of information in the field of technical standards and regulations. Off J Eur Comm L 204:37–48
9. European Parliament and Council (2001) Directive 2001/95/EC of the European Parliament and of the Council of 3 December 2001 on general product safety. Off J Eur Union L 11:4–17
10. European Parliament and Council (2002) Regulation (EC) No 178/2002 of the European Parliament and of the Council of 28 January 2002 laying down the general principles and requirements of food law, establishing the European Food Safety Authority and laying down procedures in matters of food safety. Off J Eur Comm L31:1–24
11. European Parliament and Council (2006) Regulation (EC) No 1907/2006 of the European Parliament and of the Council of 18 December 2006 concerning the Registration, Evaluation, Authorisation and Restriction of Chemicals (REACH), establishing a European Chemicals Agency, amending Directive 1999/45/EC and repealing Council Regulation (EEC) No 793/93 and Commission Regulation (EC) No 1488/94 as well as Council Directive 76/769/EEC and Commission Directives 91/155/EEC, 93/67/EEC, 93/105/EC and 2000/21/EC. Off J Eur Union L 396:1–849
12. European Parliament and Council (2011) Regulation (EU) No 182/2011 of the European Parliament and of the Council of 16 February 2011 laying down the rules and general principles concerning mechanisms for control by Member States of the Commission's exercise of implementing powers. Off J Eur Union L 55:13–18
13. Giraud G, Halawany R (2006) Consumers' perception of food traceability in Europe. Paper presented at the 98th European Association of Agricultural Economists (EAAE) Seminar, Chania, 29 June-2 July 2006. http://purl.umn.edu/10047. Accessed 03 June 2014

14. Hanson BT (1998) What happened to fortress Europe?: External trade policy liberalization in the European Union. Int Organ 52(1):55–85. doi:10.1162/002081898550554
15. Hofmann H (2009) Legislation, delegation and implementation under the Treaty of Lisbon: typology meets reality. Eur Law J 15(4):482–505. doi:10.1111/j.1468-0386.2009.00474.x
16. Houghton JR, Rowe G, Frewer LJ, Van Kleef E, Chryssochoidis G, Kehagia O, Korzen-Bohrd S, Lassend J, Pfenninge U, Strada A (2008) The quality of food risk management in Europe: perspectives and priorities. Food Policy 33(1):13–26. doi:10.1016/j.foodpol.2007.05.001
17. Levidow L, Carr S, Wield D (2005) European Union regulation of agri-biotechnology: precautionary links between science, expertise and policy. Sci Public Policy 32(4):261–276. doi:10.3152/147154305781779452
18. Marvin HJ, Kleter GA, Prandini A, Dekkers S, Bolton DJ (2009) Early identification systems for emerging foodborne hazards. Food Chem Toxicol 47(5):915–926. doi:10.1016/j.fct.2007.12.021
19. Parisi S (2012) Food packaging and food alterations. The user-oriented approach. Smithers Rapra Technology, Shawbury
20. Restuccia D, Spizzirri UG, Parisi OI, Cirillo G, Curcio M, Iemma F, Puoci F, Vinci G, Picci N (2010) New EU regulation aspects and global market of active and intelligent packaging for food industry applications. Food Control 21(11):1425–1435. doi:10.1016/j.foodcont.2010.04.028
21. Rousselle C, Pernelet-Joly V, Mourton-Gilles C, Lepoittevin JP, Vincent R, Lefranc A, Garnier R (2013) Risk assessment of dimethylfumarate residues in dwellings following contamination by treated furniture. Risk An 34(5):879–888. doi:10.1111/risa.12147
22. Rydin S (2013) Chemicals in leather: international trends on risk-based control and management. In: Bilitewski B, Darbra RM, Barceló D (eds) Global risk-based management of chemical additives II: risk-based assessment and management strategies, Hdb Env Chem. Springer, Berlin, pp 245–262. doi: 10.1007/698_2012_201
23. Xu H, Ning H, Chen Y, Fan H, Shi B (2013) Sulfanilamide-conjugated polyurethane coating with enzymatically-switchable antimicrobial capability for leather finishing. Prog Org Coat 76(5):924–934. doi:10.1016/j.porgcoat.2013.02.013

Chapter 2
EU Regulations on Chemicals in Foods

Abstract The use of chemicals in the modern society is well known. With specific reference to food productions, plant protection products and veterinary medicines are pharmacologically active substances used to fight pests and animal diseases. Food additives prolong the shelf life of foodstuffs; colours and flavourings make foods more attractive than usual. The safety and the hygiene of food products are assured by containers that are constituted of chemical substances such as plastics, metals, glass or paper. On the other hand, a number of chemicals are present in the environment as pollutants; these contaminants may be unintentionally present in raw materials used for the production and the distribution of foods. At the same time, chemicals may damage the human health, the animal or plant safety and the environment. As a consequence, the detection of harmful- or questionable-chemicals in foods and feeds has to be considered a matter of public health. Modern regulatory patterns are based on the prevention and the management of risks: food laws in the European Union share the same ratio. This Chapter is mainly focused on food-related legislations and regulations with reference to the intentional use and the accidental detection of regulated substances in foodstuffs within the EU boundaries. In addition, a Section about chemical hazards and the crisis management in the European Union is available.

Keywords European union · European food safety authority · Food additive · Food business operator · General food law · Heavy metal · Maximum level · Maximum residue limit · Mycotoxin

2.1 EU Regulations on Chemicals in Foods. General Overview

The use of chemicals in food processing is long standing. Plant protection products and veterinary medicines are pharmacologically active substances used to fight pests and animal diseases. Food additives prolong the shelf life of

© The Author(s) 2014
D. Pisanello, *Chemistry of Foods: EU Legal and Regulatory Approaches*,
SpringerBriefs in Chemistry of Foods, DOI 10.1007/978-3-319-03434-8_2

foodstuffs; colours and flavourings can be used to make attractive foods. Food safety and hygiene can be assured by means of containers that are made of chemical substances such as plastics, metals, glass or paper. On the other hand, a number of chemical substances are present in the environment as pollutants. These contaminants can be unintentionally present in raw materials for food production and distribution: the presence of similar contaminants may be not fully avoided.

At the same time, chemicals may have some impact on the human health. Since the food regulation aims at the establishment of a balance between risks and benefits of substances, the European Union (EU) legislation must meet high levels of the consumer protection as requested by the 'Treaty on the Functioning of the European Union' (TFEU), Article 168 (Sect. 1.2). Food regulatory measures must be addressed consistently with the so-called 'risk analysis' as ruled by the framework Regulation (EC) No 178/2002 (hereinafter, General Food Law, or GFL), Article 6 [55].

Chemical substances in food are covered by a number of EU legislations. These protocols can be subdivided into the following areas:

- Legislation on food a]dditives. In accordance with these documents, only explicitly authorised additives may be used, often in limited quantities when speaking of specific foodstuffs. Prior to their authorisation by the Commission, food additives have to be evaluated for their safety
- Legislation on flavourings. Clear limits are defined with reference to the presence of undesirable compounds, while a vast safety evaluation programme has been launched for flavouring substances. Only compounds with a positive evaluation can be authorised for the use in foodstuffs by means of an EU list of flavourings, in accordance with the Regulation (EU) No 873/2012. Transitional measures for other flavourings are also put in place according to the Regulation (EC) No 1334/2008
- Legislation on contaminants. The main principle is that contaminant levels shall be kept at the lowest level on condition that good working practices are established, implemented and useful. Maximum levels have been set for certain contaminants (e.g. mycotoxins, dioxins, metals, nitrates, and chloropropanols) in order to protect the public health
- Legislation on residues. This section of laws concerns veterinary medicinal products and plant protection products (pesticides) with the need of scientific evaluation before these products are authorised. If necessary, maximum residue limits (MRL) are established and in some cases the use of substances is prohibited.
- Legislation on approved food-contact materials. Basically, these objects cannot transfer their components into foods in quantities that could endanger the human health or change the composition, the taste or the texture of packaged foods.

2.2 Regulating Chemicals in Additives and Flavourings

Chemicals may preserve, impart colours or stabilise foods during production, packaging and storage steps. Enzymes have specific biochemical actions with demonstrable technological purposes at any stage of the food chain. Flavourings can give or change the odour or the taste to foods.

The EU legislation on food additives, food enzymes and food flavourings have been revised by a set of regulations approved on 2008. All these chemical can be also named 'food improvement agents'. The 'Food Improvement Agent Package' (FIAP) includes:

(1) The Regulation (EC) No 1331/2008 of the European Parliament and of the Council of 16 December 2008 establishing a common authorisation procedure for food additives, food enzymes and food flavourings [65]
(2) The Regulation (EC) No 1332/2008 on food enzymes [66]
(3) The Regulation (EC) No 1333/2008 on food additives [67]
(4) The Regulation (EC) No 1334/2008 on flavourings and certain food ingredients with flavouring properties [68].

With relation to the use of substances as additives, enzymes or flavouring, the FIAP has introduced a common assessment and a unique authorisation procedure. The effectiveness and the transparency of this procedure have to be guaranteed in a restricted temporal limit with the aim of facilitating the free movement within the Community market.

In accordance with the framework for risk assessment in matters of food safety established by the GFL, the above mentioned authorisation has to be made on the basis of an independent scientific assessment of possible health risks for the human being. This assessment must be carried out under the responsibility of the European Food Safety Authority (EFSA). In addition, a risk management decision has to be taken by the Commission under a regulatory procedure that ensures the close cooperation between the Commission and the Member States.

Authorised substances are inserted in a list—the so-called 'Union List'—which is maintained and published by the Commission. Sectoral provisions may grant the protection of scientific data and other information submitted by the applicant for a certain period of time. However, information relating to the safety of a substance cannot be regarded as confidential in any circumstances, including toxicological studies, other safety studies and raw data [65].

The right to request an updating of the Union list is granted to the Commission, an EU country or an interested party; all these entities are entitled to request the addition of a new substance to the Union List. The same right is guaranteed with relation to the removal of a substance or the addition, removal or modification of conditions, specifications or restrictions related to the presence of a substance.

As provided for by Article 9 of the Regulation (EC) No 1331/2008, the Commission has adopted the Regulation (EU) No 234/2011. This document has implemented the Regulation (EC) No 1331/2008, later amended by the Commission Implementing Regulation (EU) No 562/2012 with regard to specific data required for risk assessment of food enzymes. The Regulation (EU) No 234/2011 concerns the content, the drafting and the presentation of applications to establish or update Union Lists of food additives, food enzymes and food flavourings, the arrangements for checking the validity of applications and the type of information that must be included in the opinion of the EFSA.

In addition, the Article 3 of the Regulation (EU) No 234/2011 requires the applicant to take into account the practical guidance on the submission of applications 'Practical guidance for applicants on the submission of applications on food additives, food enzymes and food flavourings' [13], made available by the website of the 'Commission on Directorate General for Health and Consumers' (DGSANCO). This Regulation takes into consideration the EFSA's scientific opinions on data requirements (Sect. 1.5) for the evaluation of:

- Food additive applications (9 July 2009)
- Food enzyme applications (23 July 2009)
- Food flavourings (19 May 2010).

All the mentioned EFSA's opinions cover the submission of an application for the use of a new food additive, enzyme of flavouring, and an application for the modification of conditions of use for already authorised food substances. In the latter case, required data for the risk assessment may not be necessary on condition 'that verifiable justification why the proposed changes do not affect the results of the existing risk assessment' [13].

Attention would be paid to the compliance with good laboratory practices as relevant under the sectoral Directive 2004/10/EC when filling an application. In addition, the applicant should demonstrate [60] that all performed tests outside the EU territory follow the Principles of 'Good Laboratory Practice' by the Organisation for Economic Co-operation and Development.

Applicants are also requested to provide the technological justification of use: with reference to food additives, it should be demonstrated that the technological effect cannot be achieved by any other economically and technologically practicable systems. The application must also address the consideration on measures to assure the safety protection: advantages and benefits for the consumer must be explained with relation to food additives.

The request (application) is received by the Commission, which first evaluates if the application relates to uses that may have an effect on the human health; in the affirmative case, the Opinion of EFSA is mandatory.

EFSA must give an opinion within 9 months of receipt of a valid application. Should additional information be considered necessary from applicants, the 9 months-period could be extended in 'duly justified cases'. In this situation, should additional information be absent within the additional period, the EFSA would have the right of finalising its opinion on the basis of existing information.

The submission of additional information on applicants' own initiative must be sent to the Authority and to the Commission too.

Usually, substances are evaluated when a detailed dossier is available. In general, this dossier is provided by the producer or a potential user of the food additive. Information must contain the chemical identification of the additive, its manufacturing process, methods of analyses, known reactions and the final form of this substance in foods, potential and proposed uses and toxicological data.

Moreover, toxicological data must contain information on metabolism, sub-chronic and chronic toxicity, carcinogenicity, genotoxicity, reproduction, developmental toxicity and—if required—other studies. Based on this data, EFSA determines the level below which the intake of the substance can be considered safe: the so-called 'acceptable daily intake' (ADI). At the same time, the EFSA estimates also whether this ADI can be exceeded on the basis of proposed uses in different foodstuffs.

Anyway, the Commission can submit a draft regulation updating the List to the Standing Committee on the Food Chain and Animal Health established by the Article 58 of the GFL within the above mentioned 9 months-period. Should the draft regulation be not in accordance with the opinion of the Authority, the Commission would explain the reasons for its decision. It has to be remembered that EFSA's opinions may be nor required.

Decisions on the removal of a substance from the List must be adopted in accordance with the regulatory procedure with scrutiny ruled by Articles 5a (1) to (4) and the Article 7 of the Decision 1999/468/EC. In addition, every decision on the addition of a substance to the List and for adding, removing or changing conditions, specifications or restrictions associated with the presence of the substance on the Community list, shall be adopted in accordance with the regulatory procedure as ruled by Articles 5a (1) to (4) and (5) (b) and Article 7 of the Decision 1999/468/EC.

On imperative grounds of urgency, the Commission may use the urgency procedure referred to in the Article 14(5) of the Decision 1999/468/EC for the removal of a substance from the List and for adding, removing or changing conditions, specifications or restrictions associated with the presence of a substance on the Community.

2.2.1   The Community List of Food Enzymes

Basically, food enzymes are covered in the EU by the Regulation (EC) No 1332/2008 [66]. Main objectives of this document have been the definition of harmonised rules for food enzymes used in foods, including processing aids and excluding enzymes used in the production of food additives or processing aids. Moreover, the creation of a detailed 'Community list of food enzymes' which perform a technological function in foods has been decided. Probably, the above mentioned list will be the most visible and concrete result of the harmonisation of different national procedures and rules in the EU. As above mentioned, the authorisation procedure is common for food additives, enzymes and flavourings in accordance with the Regulation (EC) No 1331/2008.

As a concrete result, the use and the placing on the EU market of food enzymes will be possible only if these compounds can be preliminarily accepted and authorised for the use. The inclusion in the Community List is possible on condition that enzymes comply with general rules for EU-approved food additives:

(a) The health of the consumer is not damaged in the concentration used (for this enzyme) and on the basis of existing scientific data
(b) The use (of this enzyme) is justified by a technological need (this would also mean that this need is not obtainable in other ways)
(c) The use (of this enzyme) does not mislead the consumer, in accordance with the Directive 2000/13/EC (labelling rules).

For these reason, the definition of 'food enzyme' should be remembered in accordance to the Article 3(2) (a) of the related Regulation [66]. In detail, food enzymes are products 'obtained from plants, animals or micro-organisms or products thereof including a product obtained by a fermentation process using micro-organisms:

• Containing one or more enzymes capable of catalysing a specific biochemical reaction; and
• Added to food for a technological purpose at any stage of the manufacturing, processing, preparation, treatment, packaging, transport or storage of foods'.

Moreover, 'food enzyme preparations' are not food enzymes. This definition is correlated to a possible mixture of one or more food enzymes 'in which substances such as food additives and/or other food ingredients are incorporated to facilitate their storage, sale, standardisation, dilution or dissolution'. This definition is provided by the Art. 3(2) (b) of the Reg. (EC) No 1332/2008.

At present, the EU list of food enzymes approved for use cannot be adopted in accordance with the recent Regulation (EC) No 234/2011. In fact, an initial 24 months-period has been established for the submission deadline (11th September 2013). After this step, received applications have to be evaluated by the EFSA. Because of the necessity of creating and publishing the Community List in a single step, risk assessments have to be carried out by the EFSA for all applications with a similar and one-step procedure. As a result, the implementation of the Community List of food enzymes is expected to be published as soon as possible. The same thing is expected for risk assessments of the EFSA.

Anyway, the List of food enzymes is expected to contain:

(a) The name of the specific enzyme
(b) Detailed specifications
(c) Declared uses for food productions
(d) Conditions of use
(e) Possible restrictions for sale
(f) Specific requirements for the transparent labelling.

2.3 Focus on Food Additives

Food additives are natural or manufactured substances, added to foods for a variety of reasons:

- Original colours lost during processing should be restored in the final product. Possible solution: food colorants
- Low-sugar products should appear sweeter than usual. Possible solution: sweeteners
- The normal food deterioration should be prevented where possible and decelerated during storage, with attention to food poisoning. Possible solution: preservatives.

With reference to food additives, the basic legislation is the Regulation (EC) No 1333/2008 as amended. This document defines these products—Article 3.2(a)—as 'any substance not normally consumed as a food in itself and not normally used as a characteristic ingredient of food, whether or not it has nutritive value, the intentional addition of which to food for a technological purpose in the manufacture, processing, preparation, treatment, packaging, transport or storage of such food results, or may be reasonably expected to result, in it or its by-products becoming directly or indirectly a component of such foods' [67].

The regulation (EC) No 1333/2008 lays down rules on food additives with detailed definitions, declared uses and conditions, correct labelling and related procedures. It also addresses:

- Technological functions of food additives (Annex I)
- The Union list of food additives approved for use in food additives and conditions of use (Annex II)
- The Union list of food additives approved for use in food additives, food enzymes and food flavourings, and their conditions of use (Annex III)
- The possible prohibition of certain categories of food additives in several traditional foods by Member States (Annex IV)
- Additives labelling information for certain food colours (Annex V).

For the purpose of the Regulation (EC) No 1333/2008, a number of substances are not considered food additives. In accordance with the Article 3.2(a), the legislation [67] does not consider following 'food additives:

(1) Monosaccharides, disaccharides or oligosaccharides and foods containing these substances used for their sweetening properties
(2) Foods, whether dried or in concentrated form, including flavourings incorporated during the manufacturing of compound foods, because of their aromatic, sapid or nutritive properties together with a secondary colouring effect
(3) Substances used in covering or coating materials, which do not form part of foods and are not intended to be consumed together with those foods
(4) Products containing pectin and derived from dried apple pomace or peel of citrus fruits or quinces, or from a mixture of them, by the action of dilute

acid followed by partial neutralisation with sodium or potassium salts (liquid pectin)

(5) Chewing gum bases

(6) White or yellow dextrin, roasted or dextrinated starch, starch modified by acid or alkali treatment, bleached starch, physically modified starch and starch treated by amylolitic enzymes

(7) Ammonium chloride

(8) Blood plasma, edible gelatin, protein hydrolysates and their salts, milk protein and gluten

(9) Amino acids and their salts other than glutamic acid, glycine, cysteine and cystine and their salts having no technological function

(10) Caseinates and casein

(11) Inulin'.

Other substances often considered as additives are regulated separately, including: flavourings falling within the scope of the Regulation (EC) No 1334/2008; food enzymes falling within the scope of the Regulation (EC) No 1332/2008; substances added to foodstuffs as nutrients, substances used for the protection of plants and plant products in conformity with European Community rules relating to plant health and chemicals used for the treatment of water for human consumption falling within the scope of the Directive 98/83/EC on the quality of water intended for human consumption.

In addition, every intentionally added substance in the processing of foods without the clear 'foodstuff' status is not covered by the Regulation (EC) No 1333/2008 on condition that the same substance remain only as residues in the final food without technological effects in the final product. All these compounds are called 'processing aids'.

On the other hand, the discrimination between a processing aid and a food additive can be difficult. The presence of similar products in the finished foodstuff is rare and involuntary. Having regard to those differences, the Commission considers that a prior authorisation scheme is not justified for processing aids, since the latter does not appear potentially harmful in the same way of additives or vitamins. The definition of a processing aid in the Regulation (EC) No 1333/2008 is 'any substance which is not consumed as a food ingredient by itself, which is intentionally used in the processing of raw materials, foods or their ingredients, to fulfil a certain technological purpose during treatment or processing and which may result in the unintentional but technically unavoidable presence of residues of the substance or its derivatives in the final product, provided that they do not present any health risk and do not have any technological effect on the finished product' [67].

Only additives for which the proposed use has been considered safe are on the Union list and can be marketed. Every decision on an additive approval must take into consideration the justification of the related use in foods:

• The use of a peculiar food additive has to represent a distinctive advantage, including one or more of the technological functions set out by legislation (see mentioned Annex I), on condition that technological needs cannot be achieved by other means that are economically and technologically practicable[1]

- Moreover, this use does not present an appreciable health risk to consumers
- At the same time, the use of the peculiar food additive does not mislead the consumer.

The EU has launched an intensive activity of re-evaluation of authorised additives which should be completed by 2020. The Commission may propose a revision of the current conditions of use of additives, including possible removals from the list, on the basis of the advice of EFSA. Anyway, authorised food additives and related conditions of use are listed in the Annex II of the Regulation (EC) No 1333/2008 under specific conditions and on the basis of the peculiar food category.

In detail, the health marking of fresh meat and other markings required on meat foods may be carried out with authorised food colours. These substances are listed in the Annex II to the Regulation (EC) No 1333/2008. The same thing can be affirmed when speaking of coloured and stamped eggshells.

In addition, specifications relating to origin, purity criteria and any other necessary information are now regulated by the Commission Regulation (EU) No 231/2012.

Only food additives included in the Community list (Annex III) may be used in food additives, in food enzymes and in food flavourings under specified conditions. This 'Union list' include also 'carriers approved for use in food additives, food enzymes, food flavourings, nutrients and their conditions of use' in accordance with the Regulation (EC) No 1331/2008.

Substances falling within the definition of 'food additives' can be used for various purposes which are listed under the 26 'technological purposes' regulated under the Annex I to the Reg. (EC) No 1333/2008. Briefly, following classes of substances are authorised when in connection with specified uses:

(1) Sweeteners. Desired result: enhancement of sweet tastes
(2) Colours. Desired result: restoration or enhancement of original colours. These substances include also (a) natural constituents of foods and natural sources and (b) preparations obtained from foods and other edible natural sources by physical and/or chemical extraction resulting in a selective extraction of the pigments relative to the nutritive or aromatic constituents
(3) Preservatives. Desired results: enhancement of the shelf-life of foods; protection against deterioration caused by pathogenic and spreading micro organisms
(4) Antioxidants. Desired results: enhancement of the shelf-life of foods; protection against deterioration caused by oxidation, such as fat rancidity and colour changes

---

[1] An useful example is the use of the basic methacrylate copolymer as a glazing agent/coating agent in solid food supplements. The Authority concluded that no safety concern is demonstrable for this additive in its opinion of 10 February 2010 [19].

(5) Carriers. Authorisation for following uses and results: dissolution, dilution, dispersion or otherwise physical modification of a food additive or flavouring, food enzyme, nutrient and/or other substance added for nutritional or physiological purposes to a food without alteration of related functions. In addition, carriers cannot exert any technological effect themselves

(6) Acids. Desired results: increase of acidity values; creation of sour tastes

(7) Acidity regulators. Desired result: alteration or control of the acidity or alkalinity of foodstuffs

(8) Anti-caking agents. Desired result: reduction of the tendency of individual particles of a foodstuff to adhere to one another

(9) Anti-foaming agents. Desired results: prevention or reduction of foaming effects

(10) Bulking agents. Desired result: increase of volumetric capacities of foodstuffs without significant contributions to available energy values

(11) Emulsifiers. Desired result: realisation of a homogenous and durable mixture of two or more immiscible phases such as oil and water in a foodstuff

(12) Emulsifying salts. Desired result: conversion of proteins into a dispersed form with the homogenous distribution of fat and other components. Use in cheese preparations and melted (processed) cheeses

(13) Firming agents. Desired result(s): enhanced rigidity or firmness (examples: fruit or vegetable products); creation of stable gels if gelling agents are also used. Uses in vegetables and fruit preparations, bakery products, cooked seafood, fermented milks, dairy products, etc.

(14) Flavour enhancers. Desired result: enhancement of the existing taste and/or odour of a foodstuff

(15) Foaming agents. Desired result: creation of a homogenous dispersion of a gaseous phase in a liquid or solid foodstuff

(16) Gelling agents. Desired result: creation of a gel structure with the consequent texture

(17) Glazing agents (including lubricants). Desired result(s): these substances may be used to give a shiny appearance to the external surface of a foodstuff. In addition, the creation of a protective coating is allowed

(18) Humectants. Desired results: prevention of excessive drying or moisture loss in foods when storage atmospheres have a low degree of humidity; dissolution of powders in aqueous media

(19) Modified starches. Authorisation for use in certain foods. These chemicals are substances obtained by one or more chemical treatments of edible starches which may have undergone a physical or enzymatic treatment. Modified starches may be acid or alkali thinned. The bleached version is also available

(20) Packaging gases. Desired results: prevention of the superficial oxidation and microbial spreading on and into foods; enhanced rigidity of packages after closure

(21) Propellants. Desired result: expulsion of foodstuffs from a container

(22) Raising agents. Desired result: increase of the volumetric capacity of a dough or batter
(23) Sequestrants. Desired result: formation of chemical complexes with metallic ions
(24) Stabilisers. Desired result: maintenance of a homogenous dispersion of two or more immiscible substances in a foodstuff; stabilisation, maintenance or enhancement of an existing colour; increase of the binding capacity of the food (example: cross-links between proteins)
(25) Thickeners. Desired result: increase of the viscosity of a foodstuff
(26) Flour treatment agents. Desired result: improvement of the baking quality.

In addition to the inclusion of a food additive in the lists of authorised food additives, conditions for the authorised use are clearly stated. In particular, the type of food and the maximum permitted level of use for a particular additive in a specified foodstuff are mentioned in the legislation. Originally, these conditions were laid down in the Annexes of the three separate Directives on sweeteners (94/35/EC), colours (94/36/EC) and miscellaneous additives (95/2/EC). At present, above mentioned provisions have been transferred from the old Directives in the Union list (Annex II, Part E) and they apply from 1st June 2013.

The use of additives is very limited in certain foodstuffs. For unprocessed foodstuffs such as milk, fresh fruit and vegetables, fresh meat and water, a few additives are authorised at present. The more a foodstuff is processed, the more additives are authorised (and used). Confectionery, savoury snacks, flavored beverages and desserts are 'highly processed' foodstuffs: for this reason, many additives and correlated uses can be authorised. All possible chemicals with recognised minimum toxicological concerns may be added in almost all processed foodstuffs. Consequently, calcium carbonate (E 170) and nitrogen (E 941) are substantially authorised in many situations. On the other side, natamycin (E 235) can only be used as preservative cheese and dried sausages have to be coated superficially (surface treatment). Similarly, certain anticaking agents such as sodium ferrocyanide (E 535) may be used only in food-grade salt and related substitutes [45].

In addition, food additives may be detected in foods because of the previous presence of the same substance in a raw material or ingredient used to produce the peculiar food. This possibility is currently named 'the carry-over principle' [12]. It should be considered that the presence of these additives is subjected to the authorisation for use in the original raw material in accordance with Article 18.1(a) of the Reg. (EC) No 1333/2008. However, the 'carry-over principle' cannot be accepted in certain foods with reference to specific food additives and food colorants. These foods are listed in the Annex II, Tables 1 and 2 of the Reg. (EC) No 1333/2008.

With reference to this book, Table 2.1 shows foods and food categories in which the presence of an additive may not be permitted by virtue of the carry over rule in accordance with the Regulation (EC) No 1333/2008, Annex II, Part A, Table 1. Moreover, Tables 2.2 and 2.3 show foods and food categories in which the presence of a food colour may not be allowed by virtue of the carry over rule

Table 2.1 Foods and food categories in which the 'carry-over' rule cannot justify the presence of additives

Foods in which the presence of an additive may not be permitted by virtue of the carry over rule in accordance with the Regulation (EC) No 1333/2008, Annex II, Part A, Table 1
Unprocessed foods (definition: Article 3) of Regulation (EC) No 1333/2008
Honey
Non-emulsified oils and fats of animal or vegetable origin
Butter
Unflavoured pasteurised and sterilised (including UHT) milk and unflavoured plain pasteurised cream (excluding reduced fat cream)
Unflavoured fermented milk products, not heat-treated after fermentation
Unflavoured buttermilk (excluding sterilised buttermilk)
Natural mineral water (definition: Directive 2009/54/EC), spring water, all other bottled or packed waters
Coffee (excluding flavoured instant coffee) and coffee extracts
Unflavoured leaf tea

The 'carry-over' rule is defined in the Article 18(1) (a) of the Regulation (EC) No 1333/2008

in accordance with the Regulation (EC) No 1333/2008, Annex II, Part A, Table 2. Table 2.2 concerns all liquid and liquid-like foods; Table 2.3 shows all solid and solid-like foods.

Secondly, the carry over rule may also be allowed if a peculiar food additive, enzyme or flavouring has been added in a specific food and following conditions are satisfied:

- This additive, enzyme or flavouring is authorised in accordance with the Regulation (EC) No 1333/2008
- The same substance has been carried over to the food via an authorised food additive, enzyme or flavouring
- The absence of technological functions can be assured for the same compound in the final food.

In addition, the 'carry-over' rule may also apply if the peculiar food additive is detected in a food which is used only as a raw material for the preparation of a compound food, on condition that the final product is compliant with the Regulation (EC) No 1333/2008 ('reverse carry-over' rule).

As above mentioned, a full re-evaluation programme of authorised additives has been launched. For this reason, several additives have been discussed recently. For instance, the use of lycopene as a food colour has been restricted by the Commission Regulation (EU) No 1129/2011. The acceptable daily intake (ADI) of 0.5 mg/kg bodyweight/day for lycopene (E 160d) has been addressed by the EFSA in January 2008.

Similarly, the Commission Regulation (EU) No 232/2012 has revised the conditions of use and the use levels for quinoline yellow (E 104), sunset yellow FCF/ Orange Yellow S and ponceau 4R (cochineal red A) because of possible consequences on children's health. The EFSA has delivered different opinions about

Table 2.2 Liquid foods and beverages, including liquid-like products, in which the 'carry-over' rule cannot justify the presence of additives

Liquid foods and beverages, including liquid-like products, in which the presence of a food colour may not be allowed by virtue of the carry over rule in accordance with the Regulation (EC) No 1333/2008, Annex II, Part A, Table 2
Unprocessed foods. Definition: Article 3 of Regulation (EC) No 1333/2008
All bottled or packed waters
Milk, full fat, semi-skimmed and skimmed milk, pasteurised, sterilised, UHT types (unflavoured products)
Chocolate milk
Fermented milk (unflavoured)
Preserved milks (definition: Council Directive 2001/114/EC), unflavoured products
Unflavoured buttermilk, Cream and cream powder
Liquid oils of animal or vegetable origin
Tomato-based sauces
Fruit juice and fruit nectar. Definition: Council Directive 2001/112/EC
Vegetable juice and vegetable nectars
Extra jam, extra jelly, and chestnut purée. Definition: Council Directive 2001/113/EC
Roasted coffee, tea, herbal and fruit infusions, chicory; extracts of tea and herbal and fruit infusions and of chicory; tea, herbal and fruit infusions and cereal preparations for infusions, as well as mixes and instant mixes of these products
Wine and other products as mentioned in the Council Regulation (EC) No 1234/2007, Annex I, Part XII
Spirit drinks. Definition: Regulation (EC) No 110/2008, Annex II, paragraphs 1–14
Spirits (preceded by the name of the fruit) obtained by maceration and distillation and London gin. Definition: Regulation (EC) No 110/2008, Annex II, paragraphs 16 and 22 respectively
Sambuca, Maraschino (also named Marrasquino or Maraskino) and Mistrà. Definition: Regulation (EC) No 110/2008, Annex II, paragraphs 38 and 39 and 43 respectively
Sangria, Clarea and Zurra. Definition: Council Regulation (EEC) No 1601/91
Wine vinegar in accordance with Regulation (EC) No 1234/2007, Annex I, Part XII
Foods for infants and young children (Directive 2009/39/EC) including foods for special medical purposes for infants and young children

The 'carry-over' rule is defined in the Article 18(1) (a) of the Regulation (EC) No 1333/2008

these substances. For instance, a reduced ADI—0.5 mg/kg bodyweight/day—has been recommended for E 104 because of safety considerations by the EFSA [18].

Accordingly to the Article 24 of the Commission Regulation (EU) No 232/2012, each food containing sunset yellow FCF (E110), carmoisine (E122), allura red (E129), tartrazine (E102), ponceau 4R (E124) and E 104 must bear on the labelling an additional information relating to the 'name or E number of the colour(s)' and the alert 'may have an adverse effect on activity and attention in children' [50].

Other exceptions concerns foods where the colour(s) has been used for the purposes of health, other marking on meat products or for stamping or decorative colouring on eggshells. In addition, beverages containing more than 1.2 % by volume of alcohol are excluded in accordance with the Commission Regulation (EU)

Table 2.3 Solid and solid-like foods in which the 'carry-over' rule cannot justify the presence of additives

Solid and solid-like foods in which the presence of a food colour may not be allowed by virtue of the carry over rule as defined in the Article 18(1)(a) of Regulation (EC) No 1333/2008 (Annex II, Part A, Table 2)
Unprocessed foods. Definition: Article 3 of Regulation (EC) No 1333/2008
Oils and fats of animal or vegetable origin
unflavoured ripened and unripened cheese (unflavoured)
Butter from sheep and goats' milk
Eggs and egg products. Definition: Regulation (EC) No 853/2004
Flour and other milled products and starches
Bread and similar products
Pasta and gnocchi
Sugar including all mono- and disaccharides
Tomato paste and canned and bottled tomatoes
Fruit, vegetables (including potatoes) and mushrooms—canned, bottled or dried; processed fruit, vegetables (including potatoes) and mushrooms
Extra jam, extra jelly, and chestnut purée. Definition: Council Directive 2001/113/EC
Crème de pruneaux
Fish, molluscs and crustaceans, meat, poultry and game as well as their preparations, excluding prepared meals containing these ingredients
Cocoa products and chocolate components in chocolate products (Directive 2000/36/EC)
Salt, salt substitutes, spices and mixtures of spices
Foods for infants and young children (Directive 2009/39/EC) including foods for special medical purposes for infants and young children
Honey (Directive 2001/110/EC)
Malt and malt products

The 'carry-over' rule is defined in the Article 18(1) (a) of the Regulation (EC) No 1333/2008

No 238/2010. This situation concerns explicitly above mentioned food colorants, including E 104.

Due to the difficulties encountered during the transfer of food additives to the new categorisation system, certain errors have been introduced and should be amended. Several examples have recently concerned different additives such as curcumine, silicon dioxides and silicates. Generally, amendments concern restrictions to the authorisation of additives in certain foods or the determination of new limits such as '*quantum satis*' when speaking of silicon dioxide or silicates (E 551, E559) in accordance with the Commission Regulation (EU) No 380/2012. Further clarifications will be needed with reference to the use of additives in certain food categories.

More recently, the EFSA's re-evaluation has interested:

- A wide number of aluminium-containing food additives. The Commission Regulation (EU) No 380/2012 has amended the Annex II to the Regulation (EC) No 1333/2008 with relation to aluminium-containing food additives, including aluminium lakes, to ensure that the tolerable weekly intake (TWI)

for aluminium is defined 1 mg/kg bodyweight/week. The Regulation (EU) No 380/2012 also makes mandatory the labelling of aluminium content in aluminium lakes not intended for sale to the final consumer

- Nitrites (E 249, E 250). These additives are used as a preservative in meat products to control the possible growth of harmful bacteria, in particular *Clostridium botulinum*. Because of the risk of the formation of nitrosamines (carcinogenic substances) in traditional meat products, the current authorisation of nitrites as food additives remains valid on condition that maximum residual limits (Annex III to the Directive 95/2/EC) are considered at the end of the production process
- Aspartame (E 951). The EFSA has published its risk assessment study about aspartame. Conclusions have clearly stated that this additive is generally safe for the most part of the human population on condition that ADI does not exceed 40 mg/kg bodyweight/day. On the other hand, a limited portion of people suffering from the medical condition phenylketonuria could be affected negatively at these levels. For this reason and the public concern relating to aspartame, the above mentioned opinion could be reconsidered again in the future [16, 20].

2.3.1 Additives and Food Labelling

In accordance with the Preamble 17 of the Regulation (EU) No 1333/2008, food additives have to comply with general labelling obligations as provided for in the Directive 2000/13/EC, now repealed by the Regulation (EU) No 1169/2011 on the provision of food information to consumers. In addition, traceability requirements are mandatory as requested by Regulations (EC) No 1829/2003 and 1830/2003 concerning the traceability and labelling of genetically modified organisms and the traceability of food and feed products produced from genetically modified organisms.

In detail, the Regulation (EU) No 1333/2008 addresses specific provisions on the labelling of food additives 'intended not for sale to the final consumer' or 'not for retail sale'. According to the Article 3 no. 18 of the GFL, the final consumer 'will not use the food as part of any food business operation or activity'. In other words, the final consumer cannot be food business operators (FBO) at the same time in accordance with Article 3 (2) of the GFL. As a result, the labelling of food additives intended for sale to the final consumer has to be compliant with the general Regulation (EU) No 1169/2011 and other relevant legislations. Otherwise, general labelling requirements would be mentioned as follows, according to the Reg. (EC) 1333/2004, Article 22:

(a) The name and/or E-number
(b) The statement 'for food' or 'restricted use in food' or a more specific reference with reference to intended uses
(c) Special conditions of storage and/or use when needed

(d) The batch or lot
(e) Detailed instructions for use (the omission is not allowed when the appropriate use of the food additive could be unclear)
(f) The name or business name and address of the manufacturer, packager or seller
(g) An indication of the maximum quantity of each component or group of components with quantitative limitations in food and/or appropriate information, including also the combined percentage. The quantitative limit may be expressed either numerically or by the quantum satis principle
(h) The net quantity
(i) The date of minimum durability or use-by-date
(j) Where necessary, information on a food additive or other substances causing allergies or intolerances according to the Reg. (EU) No 1169/2011, Article 21 and Annex II.

Other requirements concern the list of ingredients in descending order, depending on the quantitative amount (percentage by weight) for mixtures of food additives with similar compounds or food ingredients. Moreover, additional information have to be labelled when speaking of food colours if listed in the Annex V to the Regulation (EC) No 1334/2008.

As above mentioned, the labelling of food additives intended for sale to the final consumer has to be compliant with the Regulation (EU) No 1169/2011 and other relevant legislations. This time, following data are mandatory:

(a) The name and E-number
(b) The statement 'for food' or 'restricted use in food' or a more specific reference with reference to intended uses.

Specific requirements are mandatory for table-top sweeteners. The name of the sweetener(s) used in the composition must be included on labels as follows: 'Name of sweetener(s)—based table-top sweetener'. In addition, the possible presence of polyols or aspartame determines additional information. When speaking of aspartame, labels must include the phrase 'contains a source of phenylalanine'. When speaking of polyols, the phrase 'excessive consumption may induce laxative effects' has to be added. This obligation concerns following substances:

• Sorbitol and sorbitol syrup (E 420)
• Mannitol (E 421)
• Isomalt (E 953)
• Maltitol and maltitol syrup (E 965)
• Lactitol (E 966)
• Xylitol (E 967).

Should food additives be listed as ingredients, they should obligatorily be mentioned in the ingredients list consistently with the Regulation (EU) No 1169/2011. As a result, every additive has to be designated by the name of the functional class followed by the specific name or EC number. For example, curcumin should be mentioned as 'colour—curcumin' or 'colour: E 100'.

2.4 Focus on Food Flavourings

Historically, flavouring components have been always extracted from natural sources. Related uses are generally correlated with pharmaceutical and cosmetic industries. Moreover, flavours can be also used as food additives.

At present, flavours can be extracted from natural sources or synthesised on a large scale. These activities started in the middle of the 19th century with two important consequences:

- The number or chemically identified flavouring substances has grown rapidly until the present time
- A new industrial sector has been progressively developed and evolved in the EU: the flavour and fragrance industry.

With reference to the position of EU flavour and fragrance industries, the predominant role of multinational companies is well known: two thirds of the current market is directly influenced by 10 multinational firms [6].

With specific reference to the food sector, the most part of flavourings are naturally present in foodstuffs. Alternatively, they can be formed during the preparation of foods. Flavourings can be added as individual chemically defined substances or as complex mixtures.

At present, flavourings and food ingredients with flavouring properties are regulated by the Regulation (EC) No 1334/2008 [68]. This document has been adopted on 16th December 2008 repealing previous laws: the Council Directive 88/388/EEC and the Commission Directive 91/71/EEC (from 20th January 2011).

In accordance with this Regulation, flavourings and source materials approved for use in and on foods have to be approved and listed in a specific 'Union list'. Moreover, specific conditions of use for these additives have to be set with additional labelling rules.

From a general viewpoint, approved flavourings and food ingredients with flavouring properties must be safe when used. Naturally, these substances have to be evaluated with a risk assessment study before of the admission in the Union List, including possible negative consequences on vulnerable groups (allergy). In addition, other factors without food safety connections should be considered including also the 'precautionary principle', the feasibility of controls and the correct information to consumers.

In accordance with the Regulation (EC) No 1334/2008, flavourings are defined in the EU as 'products not intended to be consumed as such, which are added to food in order to impart or modify odour and/or taste, which are made or consisting of the following categories: flavouring substances, flavouring preparations, thermal process flavourings, smoke flavourings, flavour precursors or other flavourings or mixtures thereof' [68].

As a result, the Regulation (EC) No 1334/2008 considers six main flavouring categories at present:

- Flavouring substances. This group includes defined chemicals with flavouring properties. Generally, flavouring substances are obtained from materials of vegetable, animal, microbiological or mineral origin. In detail, specific rules are required when speaking of a previously defined sub-category, 'natural flavouring substances'. These chemicals are 'obtained by appropriate physical, enzymatic or microbiological processes from material of vegetable, animal or microbiological origin either in the raw state or after processing for human consumption by one or more of the traditional food preparation processes listed in Annex II' of the Reg. (EC) No 1334/2008

- Flavouring preparations. These substances—essential oils from citrus fruits, peppermint or spices, extracts and tinctures, distillates from fruits, vegetables, herbs or spices—are obtained from:
  - Foods by means of appropriate physical, enzymatic or microbiological processes by one or more of the traditional food preparation processes listed in the Annex II
  - Materials of vegetable, animal or microbiological origin, other than food, by means of appropriate physical, enzymatic or microbiological processes by one or more of the traditional food preparation processes listed in the Annex II.

- Thermal process flavourings. These products are obtained from complex or simple mixtures after thermal processes. Constituents of these mixtures may be also flavourings but the use of non-flavouring compounds or ingredients is not excluded. Anyway, ingredients may be food or non-food substances. One component at least has to contain nitrogen (as proteins, amino acids and related salts, hydrolysed proteins, etc.). In addition, one component at least has to be a reducing sugar: xylose, glucose, etc.
- Smoke flavourings. Basically, these 'smoking aromas' are obtained by a complex process. In detail, primary smoke condensates, primary tar fractions and/or derived smoke flavourings as defined in the Article 3 of the Regulation (EC) No 2065/2003, are condensed with the aim of removing dangerous substances for human health [57–59]. Subsequently, the obtained 'condensed smoke' is fractionated and purified. In addition, other sub-processes may be used: absorption or membrane separation and addition of food ingredients, other flavourings, food additives or solvents [73]
- Flavour precursors. These substances may not have necessarily flavouring properties. However, their addition to foods may determine the production of peculiar flavours by means of reactions with other components during food processing. Should this aim the only purpose, these substances would be fully authorised. Main examples are oligopeptides, amino acids or carbohydrates. They can be obtained from foods and other source materials. It should be also considered that flavour precursors may belong to more than one flavouring category depending on their intended use. For instance, amino acids are flavouring substances according to the Union List, Part A; on the other hand, they may also act as precursors
- Other flavouring substances. All remaining substances—simple chemicals, foods, food mixtures, etc.—have the ability of imparting odours and/or tastes to foods.

The Regulation (EC) No 1334/2008 covers flavourings and 'food ingredients with flavouring properties', including also foods containing flavouring and/or food ingredients with flavouring properties.

In addition, this Regulation also applies to 'source materials': vegetable, animal, microbiological or mineral materials. As a result, these sources of flavourings might be not edible.

On the opposite hand, the Regulation (EC) 1334/2008 does not cover (Article 2):

- Sweeteners, sour agents or salty substances
- Raw foods (without clear flavouring properties)
- Non-compound foods and mixtures.

Anyway, the inclusion of a peculiar flavouring in the above mentioned Union List is allowed if:

- The examined substance does not pose safety risks to the health of the consumer, and
- Consumers are not misled because of the use of this substance in foods.

Should the Commission or a Member State or the Authority express doubts concerning the safety of these flavourings or food ingredients with flavouring properties, a risk assessment of such products could be carried out by the EFSA consistently with Articles 4, 5 and 6 of the Regulation (EC) No 1331/2008 [65]. Risk assessment studies should consider intake levels, the absorption, the metabolism and the toxicity of individual substances. Above all, the potential genotoxicity of flavouring substances is studied by the EFSA. This assessment is apparently the main difference between this Authority and the Joint FAO/WHO Expert Committee on Food Additive: the last Organisation has a similar role with relation to the most part of flavouring substances.

Anyway, should the EFSA identify data gaps, the request of additional data to manufacturers could be needed. After the subsequent re-evaluation, flavouring substances may be included in the Union List. At present, the database on food flavourings (web site: https://webgate.ec.europa.eu/sanco_foods/main/?event=display) can be very useful. It contains approved and flavouring substances which may currently remain on the market until the risk assessment and authorisation procedures have been concluded. With reference to excluded flavourings, these substances will be banned after an 18-months phasing-out period in accordance with the new Regulation (EU) No 872/2012. This document has adopted the Union List. For this reason, only authorised flavourings may be used after the 22th April 2013, while 18 additional months are allowed as a precautionary and phasing-out period for 'banned' flavourings.

2.4.1 Regulation on Smoke Flavourings

As mentioned in Sect. 2.4, smoke flavourings used or intended for use in or on the surface of foods are subjected to specific requirements in accordance with the

Regulation (EC) No 2065/2003 [57]. This Regulation lays down a procedure for the safety assessment and the approval of smoke flavourings—these substances are one of the six flavouring categories (Sect. 2.4)—and aims to establish a list of authorised primary smoke condensates and primary tar fractions. Following products can correspond to the definition of 'smoke flavouring' [57]:

- Primary smoke condensate. This term means the purified water-based part of the condensed smoke. For this reason, it falls within the definition of 'smoke flavouring'
- Primary tar fraction. This term means the purified fraction of the water-insoluble high density tar phase of the condensed smoke. For this reason, it falls within the definition of 'smoke flavouring'
- Primary products. Basically, these substances are primary smoke condensates and primary tar fractions
- Derived smoke flavourings. These substances or mixtures are obtained by the further processing of primary products. They can be used or intended to be used in or on foods in order to impart smoke flavours to those foods. Substantially, derived smoke flavourings are by-products.

According to Article 4.1 of the Regulation (EC) No 2065/2003, the authorisation of smoke flavourings may be subjected to specific conditions (Sect. 2.4) in spite of the classification as 'flavourings'. Moreover, authorisations should be granted for 10 years and would be renewable (Articles 9.3 and 12). This authorisation process is quite different from the general procedure for flavourings (Sect. 2.4)

By the technical viewpoint, the scientific literature seems to highlight the role of foods that are traditionally smoked: cooked smoked sausage, bacon, etc. On the other side, smoked edible products without traditional methods—crisps, soups, sauces—do not appear to affect significantly the exposure to substances such as polycyclic aromatic hydrocarbons (PAH) [86]. Basically, the use of smoke flavourings appears 'safer' than the use of smoke by burning wood or by heating saw dust or small wood chips. The reason should be substantially correlated to fractionation and purification procedures of condensed smokes.

At present, the Scientific Panel on Food Contact Materials, Enzymes, Flavourings and Processing Aids (CEF) of EFSA has completed the review of the safety of 10 primary products. On these bases and in accordance with scientific studies, the Commission has adopted the Union List of authorised smoke flavouring for use as such in or on foods applications (meat, fish and dairy products above all) and/or for the production of derived smoke flavourings. This list, published in the Commission Implementing Regulation (EU) No 1321/2013, has effect from 1st January 2014.

At present, the list contains 10 different smoke flavourings with the detailed description and the chemical characterisation of the product (including purity criteria), conditions of use and maximum allowed levels (g per kg of product). With reference to safety concerns about genotoxicity, safety assessments of the EFSA have concluded that the use of three smoke flavourings do not appear to pose safety concerns when correlated to proposed uses and use levels. On the contrary, remaining products have been re-evaluated with the conclusion that proposed uses and use levels have been judged to pose safety concerns. As a result, the CEF has considered a revision of these levels. This is the main difference between above mentioned products [53].

2.4.2   Flavourings and Food Labelling

As above mentioned (Sect. 2.4), one of main purposes of the Regulation (EC) No 1334/2008 concerns the appropriate labelling of flavourings. These substances should not be used in a way as to mislead the consumer about issues related to the nature, the freshness, the perceived quality of ingredients, the 'loyalty' or naturalness of the product (or productive processes), including also the nutritional quality [68].

Actually, flavourings remain subjected to general labelling obligations in accordance with the recent Regulation (EU) No 1169/2011 (Sect. 2.3). In addition, the Regulation (EC) No 1334/2008 contains specific provisions with relation to the labelling of flavourings sold to FBO and/or the final consumer. Sector-specific norms may apply in some cases: for instance, the traceability and the labelling of genetically modified organisms (GMO) and the traceability of food and feed products produced from GMO are covered by Regulations (EC) No 1829/2003 and 1830/2003 [58–59].

Anyway, the Regulation (EC) No 1333/2008 establishes three different set of labelling, depending on the final destination:

(a) Flavourings sold to FBO (manufacturers)
(b) Flavourings sold to final consumers
(c) Flavourings present in compound foods intended for final consumers.

'Business to business' labelling requirements for flavourings are now stipulated in Articles 15 and 16 of the Reg. (EC) No 1334/2008. Basically, the so-called 'sales description' has to be mentioned with the inclusion of:

- The minimum durability or use-by-date
- Allergen information according to the applicable legislation, and
- Special conditions for storage and/or use (when necessary).

These mandatory information have to be printed on accompanying documents and appropriate labels on packages. Moreover, Article 15 of the Reg. (EC) No 1334/2008 states clearly that all flavourings not intended for sale to the final consumer must be packaged with following information:

- The sales description, including the word 'flavouring' or a more specific name or description of the flavouring
- The declaration. Available possibilities are 'for food' or 'restricted use in food', except for more specific reference to intended food use
- The special conditions for storage and/or use (when necessary)

- The identification of lot or batch numbers
- The name or business name and address of the FBO
- The net quantity
- The date of minimum durability or use-by-date
- Information about flavourings or similar substances as defined in the Annex II to Reg. (EU) No 1169/2011 with risks of possible allergies or intolerances.

In addition, following information:

- The list of flavouring categories (where present) and other substances or materials in the product (the E-number can be used where appropriate)
- The indication of the maximum quantity of each component (or group) where a sort of quantitative limitation in foods is known may be placed on accompanying documents at present, on condition that the indication 'not for retail sale' is easily visible on containers.

It has to be remembered that flavourings for retail sale have to comply with the Reg. (EC) No 1334/2008, Article 17, and the general Regulation (EU) No 1169/2011, including the Annex V (exemption from the requirement of the mandatory nutrition declaration).

With reference to the labelling, other relevant norms for flavourings are:

- The Directive 2011/91/EU on indications or marks identifying the lot to which a foodstuff belongs
- The Regulation (EC) No 1829/2003 on genetically modified food and feed.

With concern to flavourings added to a compound food, the list of ingredients must consider the role of flavourings in accordance with the Reg. (EU) No 1169/2011, Annex VII, Part D, from 13th December 2014.

Finally, flavourings have to be named 'flavouring(s)' or labelled with a more specific name or description if one or more of flavouring components are defined according to Article 3(2) of the Regulation (EC) No 1334/2008, points (b) to (h). Interestingly, there is a specific norm for quinine and/or caffeine when used as flavourings in the production or preparation of a food. These chemical substances have to be mentioned with the exact name in the list of ingredients immediately after the term 'flavouring(s)'.

2.4.3 Flavourings and Food Labelling: Natural or Artificial Compounds?

With reference to flavourings, one of the most controversial points in the 'old' EU Legislation—before of the Reg. (EC) No 1334/2008—had concerned the distinction between 'natural identical' and 'artificial flavouring substances' since it was deemed misleading to the consumer. Recently, the new legislation has deleted the category of 'natural identical' substances. As a result (Article 16), the use of the term 'natural' can be currently used only if the peculiar flavouring component is obtained exclusively from natural flavouring substances or flavouring preparations.

In detail, 'natural flavours' should be obtained by flavouring components of natural origin and the source of flavourings should be labelled with one exception: omissions may be allowed if intended source materials are not be recognisable in the flavour or the taste of foods. Anyway, the simple mention of natural sources obliges manufacturers to 'link' 95 % at least of flavouring components to this source. On the other hand, the remaining 5 % may be derived from other natural origins on condition that flavours are easily recognisable (Articles 16.4 and 16.5). Anyway, all declared 'natural' flavourings, with or without relation to the 'main' source material (95 % of flavouring components) have to be easily recognised and the major contributor (by weight) needs to be mentioned first.

Should the source be not easily recognisable, the peculiar product should be named and labelled 'natural flavouring' (Article 16.6).

It should also be considered that definitions 'smoke flavouring(s)', or 'smoke flavouring(s) produced from food(s) or food category or source(s)' are compulsory if:

- The composition of flavouring substances contains one or more flavourings as defined in point (f) of Article 3(2) of Regulation (EC) No 1334/2008, and
- This flavouring gives a smoky flavour to the food.

2.5 Regulatory Framework on Contaminants

Basically, contaminants are substances that may affect human health and well-being. Legally speaking, contaminants are defined as 'substances that have not been intentionally added to food but that may be present in food as a result of the various stages of its production, packaging, transport or holding. They also might result from environmental contamination' [8].

Because of the connection between the detection of contaminants in foods, the quality of edible goods and food safety risks, the EU legislation has stated clearly that contaminated foods exceeding defined limits for specific contaminants cannot be put on the market: with relation to safety risks, the toxicological viewpoint is mainly considered. The Regulation (EC) No 1881/2006, Article 1(1) allows the circulation on the EU market of contaminated foodstuffs listed in the related Annex on condition that detected contaminants do not exceed ML; otherwise, contaminated foods cannot be placed on the market [35]. Moreover, this requirement is correlated with the concept of 'unsafety' under the rule of Article 14 of the GFL [55].

In addition, the main difficulty with contaminants is the number of these substances. Because of the occurrence of these chemicals in the environments and the consequent impossibility to impose a total ban, the best strategy has appeared the quantitative limitation of all contaminants as low as possible. As a consequence, the EU has progressively established clear 'maximum levels' (ML) for most dangerous contaminants. Substantially, the risk evaluation is determined on the basis of known toxicological concerns and the potential presence in the food chain.

At present, following categories of contaminants have been examined and taken into account (Sect. 2.5.2):

- Mycotoxins (aflatoxins, ochratoxin A, *Fusarium* toxins, patulin)
- Metals (cadmium, lead, mercury, inorganic tin)
- Dioxins and polychlorinated biphenyls (PCB)
- PAH
- 3-monochloropropane-1,2-diol (3-MCPD)
- Melamine and its structural analogues
- Nitrates.

First of all, ML can be set with reference to two reliable criteria:

(a) Every contaminant should be eradicated in foods, but this desire is practically unrealisable. On the contrary, ML should be set taking into account the real presence in the environment and 'Good Manufacturing Practices' (GMP) in the agricultural, fishery and other food processing sectors. In addition, the risk related to the consumption of the food has to be considered
(b) ML may be also achievable on two 'levels: the real ML value and the 'lowest' ML for a specific contaminant. With concern to the last possibility, the careful selection of raw materials is necessary. In addition, lowest ML should be addressed for the health protection of vulnerable consumers such as infants and young children.

Anyway, ML values are decided on the basis of scientific advices. The EFSA has this role at present in the EU. As above mentioned, toxicological concerns are extremely important. Moreover, ML should be extremely low and reasonably achievable when:

- Examined contaminants are genotoxic carcinogens, and
- The current exposure of the population or restricted vulnerable groups may be equal or higher than tolerable daily intake (TDI) values for a specific contaminant.

At present, main EU norms on contaminants are represented by:

- The Council Regulation (EEC) No 315/93 of 8 February 1993 laying down Community procedures for contaminants in food [8], by the European Economic Community (EEC)
- The Commission Regulation (EC) No 1881/2006 of 19 December 2006 setting maximum levels for certain contaminants in foodstuffs [35].

Basically, a premise should be made with concern to the sampling of foods. Interestingly, Member State authorities are fully responsible for this activity with the aim of assuring the compliance with the legislation.[2] On the other

---

[2] For instance, the Commission Regulation (EU) No 1258/2011 of 2 December 2011, amending Regulation (EC) No 1881/2006 as regards maximum levels for nitrates in foodstuffs, has recently confirmed the main role of Member States when speaking of monitoring activities with relation to nitrate levels in suspected vegetables, in particular green leaf vegetables [45–49].

hand, the responsibility for imported foodstuffs and the necessary compliance with the EU legislation is completely recognised to 'the country of origin', while imported goods are subsequently controlled at EU borders and on the market.

2.5.1 Setting ML

Basically, the Reg. (ECC) No 315/93 corresponds to a common approach to food contaminants on the procedural level: ML have to be set in a second step. Anyway, should a determined contaminant be covered by more specific Community rules, the Reg. (EEC) No 315/93 would not be applied to this situation.

Main requirements of this legislation can be summarised here as follows Reg. (ECC) No 315/93:

(a) FBO are fully responsible for keeping contaminant levels as low as possible by means of good practices in accordance with the Article 3(2)
(b) Should a specific contaminant be discussed with reference to the necessity of setting up a ML value in foods, the Commission would adopt a regulation with proposed ML in accordance with the regulatory procedure with scrutiny referred to in Article 8(3). This 'Comitology' system has been briefly discussed in Sect. 1.2, including the appeal to a specific committee.

Should a specific ML be set, Member States could not prohibit, restrict, or impede the placing on the market of 'compliant' foods in the EU market. On the other hand, should ML be absent in the EU legislation, every Member State could apply the relevant national provision. This means also that all national norms governing contaminants in foods have to be notified to the Commission and the other Member States with the aim of blocking the free trade on the EU market. Envisaged national measures may be applied by the Member States only three months after the notification has been made and on condition that the Commission's opinion is not negative.

2.5.2 Enforcing Legislation on Contaminants

The legal definition of EU-harmonized ML for contaminants does not appear sufficient: the enforcement and the implementation (sampling procedures, rules on reporting and interpretation, etc.) play a vital role. With relation to the Commission Regulation EC No 1881/2006 of 19 December 2006 setting maximum levels for certain contaminants in foodstuffs, this document has entered into force on 1st March 2007 [35].

At present, specific ML in certain foods have been defined for the following categories of contaminants:

- Mycotoxins (aflatoxins, ochratoxin A, patulin, deoxynivalenol, zearalenone, fumonisins, T-2 and HT-2 toxin)
- Metals (cadmium, lead, mercury, inorganic tin)
- Dioxins and dioxin-like PCB
- PAH such as benzo[a]pyrene
- Melamine and its structural analogues
- 3-MCPD
- Nitrates.

Some detailed instruction should be remembered here with relation to analytical results.

First of all, the Article 3 states that compliant foodstuffs cannot be mixed with non-compliant foodstuffs. This requirement might mean also that analytical results have to be free from matrix-related errors and interferences, and they have to be correlated with exactly determined foodstuffs (lot or batch, durability or use by date, etc.).

Moreover, the Article 8 stipulates that sampling methods and analytical procedures and results concerning dioxins and dioxin-like PCB have to comply with the Commission Regulation (EC) No 1883/2006 [36].

Generally, ML have to be examined and set with concern to the edible part of foodstuffs except for possible specifications according to Section 3 of the Annex of Regulation (EC) No 1881/2006 [35]. In addition, modifications of the concentration of the contaminants have to be taken into account when speaking of dried, diluted, processed and compound foodstuffs. The last situation is particularly important because of the necessity of considering relative proportions in the list of ingredients.

Following Sections are dedicated to specific ML.

2.5.2.1 Maximum Levels in Foodstuffs for Nitrates

The word 'nitrate' means a group of chemical compounds where the nitrate ion is the common presence. Chemically, 'nitrate' is defined as a natural compound [23]; the presence of the anion nitrate in the nitrogen cycle is well known as affirmed by the Panel on Contaminants in the Food Chain (CONTAM). In detail, it is obtained from ammonia by means of nitrifying bacteria with an intermediate step (the production of the anion nitrite). In addition, nitrate can be converted in gaseous nitrogen by means of the action of denitrifying bacteria [74]. Finally, 'nitrate' is an approved food additive [23]. Because of the related role in the nutrition and functional activities of vegetables, it can be notably accumulated in plants. It plays an important role in the nutrition and the function of plants.

By the toxicological viewpoint, nitrate is not recognised as a toxic substance. On the other hand, related metabolites and reaction products such as nitrite,

N-nitroso compounds and nitric oxide are suspect molecules because of their role in methaemoglobinaemia and carcinogenesis [23].

As a result, the Commission Regulation (EU) No 1258/2011 of 2 December 2011 has amended recently the main Regulation (EC) No 1881/2006 with exclusive relation to nitrates and related ML in foodstuffs. In detail, Member States are now forced to monitor nitrate levels in 'suspected' vegetables such as green leaf vegetables. In addition, results have to be communicated regularly to the EFSA [46].

With reference to the current ML for nitrates, the recent Opinion of the CONTAM Panel has confirmed the ADI for nitrates of 3.7 mg/kg bodyweight/day [23]. This amount corresponds to 222 mg of nitrate per day for a 60 kg-adult. The Opinion has taken into account possible negative effects and positive advantages of the ingestion of nitrate when contained in vegetables.

In addition, the Commission Regulation (EU) No 1258/2011 has re-defined ML for nitrates in certain vegetables and other products:

- Fresh spinach (*Spinacia oleracea*)
- Preserved, deep-frozen or frozen spinach
- Protected and open-grown fresh Lettuce (*Lactuca sativa L.*)
- 'Iceberg' type lettuce
- Rucola (*Eruca sativa, Diplotaxis* sp., *Brassica tenuifolia, Sisymbrium tenuifolium*)
- Processed cereal-based foods and baby foods for infants and young children.

The last situation is particular because established ML values refer to ready-to-use foods, including reconstituted products. However, the CONTAM Panel has also concluded that the exposure to nitrate is unlikely to be a health concern at present when speaking of vegetable products, although some risk for infants eating more than one spinach meal per day cannot be excluded. This conclusion has been published in a subsequent Opinion [26]. In addition, the EFSA has considered that the modification of nitrate contents due to the processing of food ties such as washing, peeling and/or cooking cannot be taken into account at present because of insufficient data. Consequently, the quantitative exposure of nitrate might be overestimated.

With reference to processed cereal-based foods and baby foods for infants and young children, the current legislation is the Commission Directive 2006/125/EC on processed cereal-based foods and baby foods for infants and young children. However, this Directive will be totally repealed from 20th July 2016 by the Regulation (EU) No 609/2013 on food intended for infants and young children, food for special medical purposes, and total diet replacement for weight control [71].

On a general level, current ML for nitrates in vegetables are laid down in the Annex, Section 1 of the Commission Regulation (EC) No 1881/2006. In detail, five food categories are considered when speaking of ML for nitrates [35, 46]:

- Fresh spinach. ML are 3,000 and 2,500 mg of nitrate/kg depending on the date of harvest
- Preserved, deep-frozen or frozen spinach. ML is 2,000 mg of nitrate/kg

- Fresh lettuce (protected and open-grown lettuce). ML are between 2,500 and 4,500 mg of nitrate/kg depending on the type of lettuce and the date of harvest
- 'Iceberg'-type lettuce. ML are between 2,000 and 2,500 mg of nitrate/kg depending on the type of lettuce
- Processed cereal-based foods and baby foods for infants and young children. ML is always 200 mg of nitrate/kg.

Three considerations have to be made about the assessment of nitrate amounts in foodstuffs according to the existing EU legislation.

First of all, some ML may vary for fresh spinach and fresh lettuce depending on the season, because of well known modifications of climatic conditions, production methods and eating habits in different EU countries. As a consequence, ML for nitrates are considered higher in harvested vegetables between the 1st October and the 31th March, if compared with seasonal ML for samples harvested between the 1st April and the 30th September.

Another remarkable difference concerns the discrimination between 'lettuce grown under cover' and 'lettuce grown in the open air'. ML values for nitrates are higher in the last category.

With concern to sampling and analytical procedures, it has to be noted that:

- The current reference text is the Commission Regulation (EC) No 1882/2006 [37]
- The number of samples is not defined according to the Commission Regulation (EC) No 1882/2006
- There are not prescribed official methods for the determination of nitrates in foodstuffs, although the Commission Regulation (EC) No 1882/2006 mentions analytical procedures with comparable levels of performance. Actually, laboratories can use each method of analysis on condition that they strictly fulfil analytical requirements laid down in the respective legislation and this diligence is demonstrable.

As a result, sampling methods are known while prescribed analytical methods for nitrates have to comply with the provisions of items 1 and 2 of Annex III (characterisation of methods of analysis) to the Regulation (EC) No 882/2004 of the European Parliament and of the Council of 29 April 2004. This is a very general Regulation and has to be considered when speaking of official controls in a large variety of foodstuffs [61].

In accordance to the Regulation (EC) No 882/2004, performance criteria for the analytical recovery and precision are set as follows:

- When the concentration is <500 mg/kg, the 'recommended' recovery value is 60–120 %
- On the other hand, should the concentration be ≥500 mg/kg, the 'recommended' recovery value should be 90–110 % [37].

In addition, the Horwitz equation has to be used for the calculation of precision values in accordance to Annex D.3.2 of the Commission Regulation (EC)

No 1882/2006. Maximum allowed values should be calculated as two times Horwitz values, when the above mentioned equation gives one value only [37]. Mathematically, the Horwitz equation is a linear function that calculates the regression line between the logarithm of the standard deviation for a single substance obtained by an interlaboratory dataset and the logarithm of the concentration of the same compound [1].

Finally, applicable analytical methods for the determination of nitrates and nitrite in various foodstuffs have already been standardised by the European Committee for Standardization (CEN). With exclusive reference to the determination of nitrate and nitrite in vegetables, vegetable products, including vegetable containing food for babies and infants as well as in meat and meat products, the standard series EN 12014, parts 1, 2, 3, 4, 5, and 7 have been made available by the technical committee CEN/TC 275 'Food analysis—horizontal methods' [37]. Anyway, the analytical result is always expresses as 'x ± U', where x is the analytical result and U is the expanded measurement uncertainty for a level of confidence of approximately 95 %. The expression of the analytical result is reported in the Commission Regulation (EC) No 1882/2006.

2.5.2.2 Maximum Levels in Foodstuffs for Mycotoxins

On a general level, current ML for mycotoxins are laid down in the Annex, Section 2 of the Commission Regulation (EC) No 1881/2006. However, these data are differentiated for aflatoxins, ochratoxin A, patulin, deoxynivalenol, zearalenone, fumonisins, T-2 and HT-2 toxins.

ML for Aflatoxins

According to the Scientific Panel on Food (SCF), the genotoxic and carcinogenic role of aflatoxins is well known [83]. At the same time, the detection of these molecules in foods and feeds is extensively reported because of the potential food contamination by moulds, especially *Aspergillus flavus* and *A. parasiticus*. Storage and environmental conditions have an important role: mould spreading is optimal in warm and humid areas. Generally, most contaminated food products are:

- Tree nuts (examples: almonds, Brazil nuts, pistachios, walnuts, etc.)
- Ground nuts
- Dried fruits
- Spices
- Crude vegetable oils
- Cocoa beans
- Maize.

In detail, ML values for aflatoxins have to be considered for the total quantity of these molecules in foods (the sum of aflatoxins B_1, B_2, G_1 and G_2) and also for the single aflatoxin B_1 content. This decision has been taken because of the high toxicity of aflatoxin B_2 if compared with remaining aflatoxins. Aflatoxin M_1 has been also considered in foods for infants and young children: a possible reduction of the current ML has been recently made. Before discussing ML for these molecules, it should be considered that these values have been decided on the basis of scientific Opinions by the EFSA and the necessity of aligning EU limits to the opinion of the Codex Alimentarius Commission.

At present, ML values for total aflatoxins range from 4.0 to 15.0 μg/kg. Actually, it should be considered that:

- Higher ML have been established for groundnuts (15 μg/kg) and nuts, maize and dried fruit (10 μg/kg) if these foodstuffs are (a) ingredients in other foods or (b) subjected to a sorting or other physical treatment before their human consumption
- ML values are not defined for total aflatoxins when speaking of raw milks, milk and milk products, processed cereal-based foods and baby foods for infants and young children, infant formulae and follow-on formulae, including infant milk and follow-on milk, dietary foods for special medical purposes intended specifically for infants.

Consequently, the 'basic' ML for total aflatoxin appears to be 4.0 μg/kg.

With relation to the single aflatoxin B_1, ML values range from 0.10 to 8.0 μg/kg. Once more, it should be considered that:

- Higher ML have been established for groundnuts (8 μg/kg) and nuts, maize and dried fruit (5 μg/kg) if these foodstuffs are (a) ingredients in other foods or (b) subjected to a sorting or other physical treatment before their human consumption
- There are not ML for total aflatoxins when speaking of raw milks, milk and milk products, infant formulae and follow-on formulae, including infant milk and follow-on milk.

As a result, the 'basic' ML for aflatoxin B_1 appears to be ranged between 0.10 μg/kg, when speaking of 'processed cereal-based foods and baby foods for infants and young children, dietary foods for special medical purposes intended specifically for infants', and 2.0 μg/kg for groundnuts, dried nuts, processed products, cereals and cereal products as intended for immediate human consumption or for use as ingredient.

Finally, ML for aflatoxin M_1 ranges between:

- 2.5×10^{-2} μg/kg when speaking of 'infant formulae and follow-on formulae' and 'dietary foods for special medical purposes intended specifically for infants' on the one hand, and
- 5.0×10^{-2} μg/kg for raw milks, heat-treated milks and milk for the manufacture of milk-based products.

In accordance with the Commission Regulation (EC) No 1881/2006, Annex, Section 2, there are not further ML for aflatoxin M_1 at present.

Methods of sampling and analysis for the official control of the levels of mycotoxins, including also ochratoxin A (section 'ML for Ochratoxin A') and patulin
(section 'Patulin'), in foodstuffs are addressed by the Commission Regulation
(EC) No 401/2006 and subsequent amendments [38]. In detail, methods of sampling for total aflatoxins and aflatoxin B_1 concerns (Annex I):

- Cereals and cereal products (Part B), including also ochratoxin A and *Fusarium*
  toxins
- Dried fruit but with the exception of dried figs (Part C)
- Dried vine fruit (Part C) with exclusive reference to ochratoxin A
- Dried figs, groundnuts and nuts (Part D), as amended by the recent Commission
  Regulation (EU) No 178/2010
- Groundnuts (peanuts), other oilseeds, apricot kernels and tree nuts, as added by
  the recent Commission Regulation (EU) No 178/2010
- Spices (Part E) with reference to ochratoxin A, aflatoxin B_1 and total aflatoxins,
  as amended by the recent Commission Regulation (EU) No 178/2010
- Milk and milk products and infant formulae and follow-on formulae, including
  infant milk and follow-on milk (Part F), with exclusive reference to aflatoxin M_1
- Roasted coffee, ground roasted coffee, soluble coffee, liquorice root and liquorice extract (Part G) with exclusive reference to ochratoxin A, as amended by
  the recent Commission Regulation (EU) No 178/2010
- Wine, grape juice and grape (Part H) with exclusive reference to ochratoxin A
- Fruit juices, fruit nectar, spirit drinks, cider and other fermented drinks derived
  from apples or containing apple juice (Part H) with exclusive reference to patulin
- Solid apple products and apple juice and solid apple products for infants and
  young children (Part H) with exclusive reference to patulin
- Dietary foods (milk and milk products) for special medical purposes intended
  specifically for infants, baby foods and processed cereal-based foods for infants
  and young children, other baby foods (Part J) with exclusive reference to total
  aflatoxins and single mycotoxins.

Finally, the recent Commission Regulation (EU) No 178/2010 has also added the
Annex III concerning official sampling procedure for vegetable oils with reference
to mycotoxins, in particular aflatoxin B_1, total aflatoxins and zearalenone.

On the other hand, analytical methods and control requirements have been
addressed in the same document, Annex II [38], as amended by the recent Commission
Regulation (EU) No 178/2010. Anyway, official controls have to be carried out as
specified in accordance with the provisions of the Regulation (EC) No 882/2004, as
stated in Annex I, Part A of the Commission Regulation (EC) No 401/2006.

All these procedures concern the official control of EU commodities. On the
other side, the official control on aflatoxins in imported commodities has to be carried out on the basis of the existing EU legislation according to Article 53 of the
GFL, including also:

- The Regulation (EC) No 1152/2009 (safeguard measures)
- The Regulation (EC) No 669/2009 (increased frequency of controls at the
  import)

- The Decision 2008/47/EC (reduced frequency of controls at the import).

With relation to the Regulation (EC) No 1152/2009, it can be remembered [41] that:

- At present, higher safety measures for entering EU market can be carried out when speaking of following imported goods: Brazil nuts in shell from Brazil, peanuts from China, peanuts from Egypt, pistachios from Iran, figs, hazelnuts and pistachios from Turkey
- All product/country combinations covered by the provisions of the Regulation (EC) No 1152/2009 are mentioned in the following list:
  - Pistachios from China
  - Groundnuts from Egypt
  - Groundnuts from China
  - Almonds from United States of America
  - Dried figs, hazelnuts and pistachios from Turkey
  - Brazil nuts in shell from Brazil
  - Derived products and compound products (containing the above mentioned food for more than 20 %)

- The FBO is always responsible for all costs resulting from the official controls, including sampling, analysis, storage and any measures taken following the possible non-compliance of imported commodities.

ML for Ochratoxin A

Ochratoxin A (OTA) is a peculiar mycotoxin. It is synthesised by specific fungi such as *Penicillium* and *Aspergillus* genera. OTA-related food contamination episodes have been extensively reported in the scientific literature. The contamination usually concerns cereals and cereal products, coffee, beers, dry vine fruits and wine, cocoa products, nuts and spices, etc. [22].

In addition, a certain degree of contamination appears unavoidable at present because OTA may initially be present in animal feeds. As a result, this mycotoxin may be found in edible offal and blood serum in a residual amount. On the other hand, OTA contamination in meat, milk and eggs appears negligible [22].

Basically, the initial EFSA assessment had established a TDI of 120 ng/kg bodyweight for OTA. Subsequently, the CONTAM Panel has confirmed proposed ML for OTA [27].

At present, the Regulation (EC) No 1881/2006 [35] mentions ML values for OTA in foodstuffs (section 'ML for Aflatoxins') such as:

- Unprocessed cereals and all products derived from unprocessed cereals
- Dried vine fruit (currants, raisins and sultanas)
- Roasted coffee beans and ground roasted coffee
- Soluble coffee (instant coffee)
- Wine (including sparkling wine, excluding liqueur wine and wine with an alcoholic strength of not less than 15 % as volume), and fruit wine

- Aromatised wine, aromatised wine-based drinks and aromatised wine-product cocktails, grape juice, concentrated grape juice as reconstituted, grape nectar, grape must and concentrated grape must as reconstituted, intended for direct human consumption
- Processed cereal-based foods and baby foods for infants and young children
- Dietary foods for special medical purposes intended specifically for infants
- Green coffee, dried fruit other than dried vine fruit
- Beer
- Cocoa and cocoa products
- Liqueur wines
- Meat products
- Spices and liquorice.

According to the above mentioned norm and the recent Commission Regulation (EU) No 105/2010 (amendment), the position of a peculiar food group appears very important in relation to OTA contamination. In fact, remarkable OTA amounts have sometimes been found in spices and liquorice. As a result, new ML values have been introduced for OTA with concern to spices and liquorice only [43].

However, it has to be observed that final ML for OTA in spices have been determined as follows:

- 30 μg/kg as from 1st July 2010 until 30th June 2012, and
- 15 μg/kg as from 1st July 2012.

Subsequently, this ML has been amended as follows:

- 30 μg/kg until 31th December 2014, and
- 15 μg/kg as from 1st January 2015.

This amendment has been necessary [51] because of the predictable impossibility of obtaining projected lower ML for OTA in *Capsicum* species (dried fruits thereof, whole or ground, including chillies, chilli powder, cayenne and paprika).

With reference to liquorice (*Glycyrrhiza glabra*, *G. inflate* and other species), proposed ML for OTA are 20 μg/kg and 80 μg/kg when speaking of 'liquorice root, ingredient for herbal infusion' and 'liquorice extract, for use in food in particular beverages and confectionary' respectively.

Methods of sampling and analysis for the official control of the OTA in foodstuffs are addressed by the Commission Regulation (EC) No 401/2006 and subsequent amendments [38]. In detail, methods of sampling for OTA concerns (Annex I):

- Cereals and cereal products (Part B),
- Dried vine fruit (Part C)
- Roasted coffee, ground roasted coffee, soluble coffee, liquorice root and liquorice extract (Part G) as amended by the recent Commission Regulation (EU) No 178/2010
- Wine, grape juice and grape (Part H)
- Dietary foods (milk and milk products) for special medical purposes intended specifically for infants, baby foods and processed cereal-based foods for infants and young children, other baby foods (Part J).

On the other hand, analytical methods and control requirements have been addressed in the same document, Annex II [38], as amended by the recent Commission Regulation (EU) No 178/2010. Anyway, official controls have to be carried out as specified in accordance with the provisions of the Regulation (EC) No 882/2004, as stated in the Annex I, Part A of the Commission Regulation (EC) No 401/2006.

Patulin

Differently from other mycotoxins, patulin is generally correlated with contaminated apple products. Actually, patulin may be also found in mouldy fruits, grains and other products [3, 77]. It is synthesised by several fungal species such as *Aspergillus*, *Byssochlamys* and *Penicillium* species [34].

At present, ML for patulin have been set by the Commission Regulation (EC) No 1425/2003 [34] amending the Regulation (EC) No 466/2001. Actually, the situation may evolve towards a definition of amended ML on the basis of new scientific and technological data. Moreover, the implementation of the Commission Recommendation 2003/598/EC of 11 August 2003 states clearly that 'a revision of ML for patulin in fruit juices, concentrated fruit juices, fruit nectars, spirit drinks, cider and other fermented drinks derived from apples' is forecasted. The above mentioned situation has been caused by (a) the initial absence of set ML for patulin, in spite of the definition of a 'provisional maximum tolerable daily intake' (PMTDI) of 0.4 μg/kg body weight, and (b) the German proposal about ML for patulin in 2003. As a result, the need of a harmonising measure has been made necessary in the EU ambit.

Consequently, ML values for patulin have been mentioned in the Annex to the Regulation (EC) No 1425/2003. In general, ML range between 10.0 ppb or μg/kg in 'apple juice, solid apple products and other baby food' and 50 ppb or μg/kg in 'fruit juices and nectar, concentrated fruit juice after reconstitution, spirit drinks, cider and other fermented drinks derived from apples or containing apple juice'.

Sampling and analytical methods for the official control of the levels of patulin in foodstuffs are addressed by the Commission Regulation (EC) No 401/2006 and subsequent amendments [38], as also shown in Sect. 2.5.2.2.

ML for Fusarium Toxins

With reference to the whole group of so-called *Fusarium* toxins, the SCF Panel has issued different Scientific Opinions in 1999. The diversification of these mycotoxins justifies the number of Opinions and correlated data, TDI above all EC Commission [35].

According to the Preamble 28 of the Reg. (EC) No 1881/2006, discussed mycotoxins and correlated TDI (μg/kg body weight) are:

- Deoxynivalenol. TDI: 1 μg/kg
- Zearalenone. Temporary TDI: 0.2 μg/kg

- Fumonisins. TDI: 2 μg/kg
- Nivalenol. Temporary TDI: 0.7 μg/kg
- T-2 and HT-2 toxins. Combined temporary TDI: 0.06 μg/kg
- Trichothecenes.

Deoxynivalenol

Deoxynivalenol (DON), also named vomitoxin (Fig. 2.1), is synthesised by fungal organisms, including *Fusarium graminearum* and *F. culmorum*. Generally, DON is mainly detected in cereals and grains. In addition, DON is often found with other *Fusarium* toxins: nivalenon, zearalenone and fumonisins. In spite of clear effects on animal health—the interaction of DON with serotonergic and dopaminergic receptors may determine vomiting—DON and its metabolites are not recognised as teratogenic and genotoxic agents at present. In addition, absorbed DON is rapidly turned into less toxic products. Consequently, the diffusion of DON and its metabolites into edible tissues, milk and eggs is not remarkable and the human safety should not be compromised when speaking of consumption of animal products [31].

Zearalenone

Zearalenone (ZON), also named F-2 toxin, is a phenolic resorcyclic acid lactone biosynthesised by produced by several *Fusarium* and *Gibberella* species, including *F. graminearum, F. culmorum, G. intricans (F. equiseti), G. moniliformis (F. verticillioides)* and *F. oxysporum* [2, 29]. It is commonly associated with food contamination when speaking of corn, other cereals—wheat, barley, sorghum and rye—and feeds: climatic conditions are extremely important [2]. The EFSA has

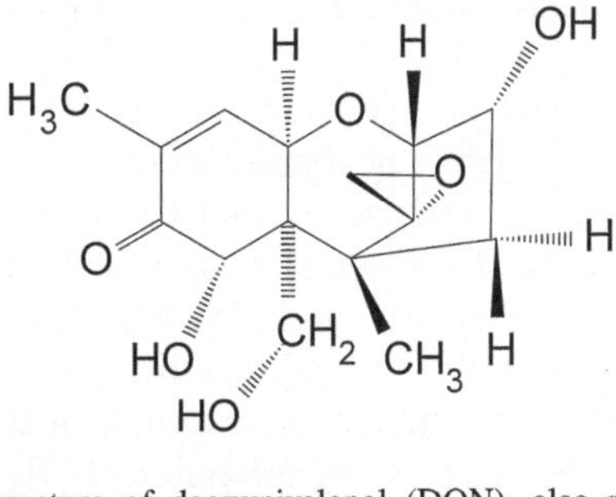

Fig. 2.1 The chemical structure of deoxynivalenol (DON), also named vomitoxin, molecular formula: $C_{15}H_{20}O_6$, molecular weight: 296.32 g/mol, Chemical Abstracts Service number: 51481-10-8. This molecule is a trichothecene synthesised by fungal organisms. Chemically, trichothecenes correspond to a heterogeneous group of tetracyclic sesquiterpenes with high thermal stability (DON is stable at 120 °C, while the tolerance is 'moderated' at 180 °C [31]. BKchem version 0.13.0, 2009 (http://bkchem.zirael.org/index.html) has been used for drawing this structure

issued in 2011 an Opinion about health risks related to the presence of ZON in foods such as breakfast cereals [29].

Because of remarkable amounts of detected ZON in several foods—wheat bran, corn, corn flour, cornflakes, grains and grain-based foods, bread and fine bakery wares, vegetable oils—the CONTAM Panel has established a TDI for ZON of 0.25 μg/kg bodyweight. The risk is substantially associated with the oestrogenic activity in sensitive animal species. According to some scientific literature, the binding of this mycotoxin to oestrogen receptors appears to be 20-fold lower if compared with the same attitude of another agent as 17 β-estradiol [2, 29].

Anyway, ML for ZON are currently mentioned in the Commission Regulation (EC) No 1881/2006, as also remembered in the recent Statement of the EFSA with relation to the evaluation of the increase of public health risks if related to possible temporary derogations from ML of DON, ZON and fumonisins for maize and maize products [17]. At present, ML for ZON range between 200 μg/kg (in processed maize-based foods and baby foods for infants and young children) and 4,000 μg/kg in unprocessed maize with the exception of unprocessed maize intended to be processed by wet milling [17].

Fumonisins

Fumonisins (FUMO) are mycotoxins produced by several fungal species. The biosynthesis of these structurally related toxins is mainly ascribed to *Fusarium verticillioides* and *F. proliferatum;* most contaminated foods seem maize and maize-based products [2, 17]. Chemically, fumonisin B_1 is the most known, prevalent and toxic within these molecules: the whole group contains at least 15 other mycotoxins: B_2, B_3, B_4, etc. [2].

In addition, FUMO are often detected with DON and ZON [17]. Once more, climatic conditions are important: the synthesis of FUMO occurs in maize under warm and dry conditions [2]. With reference to risks, fumonisin B_1 is known as a carcinogenic agent in rodents; on the other hand, a significant genotoxic activity is possible [32].

At present, available data on carry-over of FUMO from animal feeds into edible tissues appear to highlight a negligible transfer. As a result, FUMO do not seem to have a remarkable influence on the total human exposure [17].

At present, ML values for FUMO are mentioned in the Commission Regulation (EC) No 1881/2006 [17]. Similarly to ZON, ML for FUMO range between 200 μg/kg (in processed maize-based foods and baby foods for infants and young children) and 4,000 μg/kg in unprocessed maize with the exception of unprocessed maize intended to be processed by wet milling [17].

T-2 and HT-2 Toxins

T-2 and HT-2 toxins are biosynthesised by different Fusarium species. At present, highest mean concentrations for T-2 and HT-2 have been found in grains and grain milling products, including mainly oats and oat products [30]. The CONTAM Panel has established a group TDI of 100 ng/kg bodyweight for the sum of T-2 and HT-2 toxins. In addition, the CONTAM Panel has considered that the possible carry-over effect of T-2 and HT-2 toxins from feeds to animal food products is not remarkable with consequent negligible effects on the human exposure [30].

However, the collection of additional data on T-2 and HT-2 in cereals and cereal products and correlated effects of food processing (i.e. cooking) is still in progress, in accordance with the recent Commission Recommendation 2013/165/EU. At the same time, Member States are encouraged to monitor the situation with reference to the collection of analytical data for T-2 and HT-2 toxins (and other *Fusarium* toxins). For these reasons, specific EU harmonised MLs for T-2 and HT-2 toxins in food and feed products have not been decided at present in the EU despite similar ML in other countries [30].

On the other hand, current ML values for *Fusarium* toxins have been progressively introduced as amendments to the Regulation (EC) No 1881/2006 for DON, ZON and FUMO in maize and maize products.

2.5.2.3 Metals

When speaking of metallic contaminants and related ML in the EU, three chemical elements at least have to be taken into account. These elements are cadmium, lead and mercury, in strict accordance with the Commission Regulation (EC) No 1881/2006 [35]. In addition, inorganic tin has to be mentioned.

On a general level, methods of sampling and analysis for the official control of above mentioned metals and other chemicals in foodstuffs are laid down in the Commission Regulation (EC) No 333/2007 [39].

Lead

With relation to lead (Pb), this element is substantially found in its inorganic form when speaking of environmental detection. For this reason, the human exposure is mainly caused by food and water consumption. Moreover, other contamination media are naturally air, dust and soil [28].

At present, the EU estimation of Pb dietary exposure ranges in the adult human are between 0.36 and 2.43 μg/kg body weight per day. On the other hand, vulnerable groups as infants and children appear be variably exposed [28]: from 0.21 to 0.94 μg/kg per day when speaking of infants only, while ranges seem to increase in children (0.8–3.10 μg/kg in average consumers, maximum exposure: 5.51 μg/kg in high consumers).

With reference to foods, cereal products seem the main contributor to the dietary Pb exposure. On the other hand, children appear to absorb more Pb than adults with augmented accumulation risks in soft tissues and bones.

For these and other reasons—risk of developmental neurotoxicity in young children, cardiovascular effects and nephrotoxicity in adults—the CONTAM Panel has concluded that the current 'provisional tolerable weekly intake' (PTWI) of 25 μg/kg bodyweight is no longer appropriate at present [28].

ML values for Pb have been originally mentioned in the Annex, Section 3 of the Commission Regulation (EC) No 1881/2006 as amended by the Commission Regulation (EU) No 420/2011. Briefly, the lowest ML of 0.02 mg/kg is set for infant formula, follow-on formula, as well as raw milk, heat-treated milk and milk for the manufacture of milk-based products. On the other hand, the highest ML of 3.0 mg/kg is considered for food supplements. Naturally, one of main problems in the determination and the assessment of safety risks is correlated to the analytical error. Consequently, it has to be remembered that analytical performance charac-teristics for Pb are mentioned in the Regulation (EC) No 333/2007.

On the other side, water is a different 'food product' because of the wide availability, the correlate meaning of 'publicly available good and service' and the easy contamination. With relation to 'drinking waters intended for human consumption', the Council Directive 98/83/EC has provided harmonised levels for Pb: every Member State has to set a ML of 25 μg/l until the 1st December 2013 and a new ML of 10 μg/l after this date. The last limit is also mentioned in the Commission Directive 2003/40/EC (list, concentration limits and label-ling requirements for constituents of natural mineral waters and conditions for using ozone-enriched air for the treatment of natural mineral and spring waters). Moreover, this Directive sets performance characteristics for the analytical deter-mination of Pb in waters: these conditions are legally different from those stated in the Regulation (EC) No 333/2007 (Pb in foodstuffs).

The third source of contamination for food products is correlated to packaging materials and objects. Detailed ML for Pb and cadmium are regulated in ceramics by the Council Directive 84/500/EEC and subsequent amendments, when speak-ing of migration from ceramic materials. According to the Council of European Communities (CEC), the analytical method for the determination of the migration of these two elements is also shown in the above mentioned Directive [5].

Finally, the possible contamination of animal feeds has to be considered. The Directive 2002/32/EC mentions ML for Pb in different feed products, on condition that the examinable level is based on a feed product with a moisture content of 12 %. In accordance with this Directive, the official analysis must be performed by carrying out a digestion of feeds in nitric acid (5 %) for 30 min at boiling tempera-ture. However, equivalent procedures can be applied on condition that the recovery efficiency of Pb in feeds is at least as good as that of the official procedure.

Cadmium

With relation to cadmium (Cd), this metal is detected and considered as food con-taminant and environmental pollutant at the same time because two contamination sources are demonstrated at present. In fact, the natural occurrence of Cd has to be considered with industrial and agricultural sources of contamination. Moreover, the importance of this metal is relevant when speaking of the 'non-smoking general population': with reference to this 'restricted' group, foodstuffs are recognised as the main source of Cd exposure. Moreover, other contamination media are air, dust and soil [25].

In accordance with the CONTAM Panel, the absorption of Cd in the human being is not important. However, this metal can remain in tissues: at present, the known biological half-life is between 10 and 30 years. Moreover, different dysfunctions are ascribed to Cd in humans: renal dysfunctions and bone demineralisation. The International Agency for Research on Cancer has also classified Cd as a group 1-human carcinogen [25].

After a specific request by the EU Commission, the CONTAM Panel has carried out a risk assessment study about connections between health dangers and the presence of Cd in foodstuffs [25]. Substantially, Cd appears more abundant in chocolate, fish and seafood, seaweed and foods for special dietary uses. However, the abundance of Cd in most contaminated foodstuffs is correlated with remarkable concentrations of the same metal in the environments. Moreover, the use of fertilisers with increased Cd amounts has to be considered.

Anyway, the mean exposure for adults in the EU is close or slightly exceeds a tolerable weekly intake (TWI) of 2.5 µg/kg bodyweight. On the other side, four different subgroups of the human population—vegetarians, children, people living in Cd-contaminated areas and smokers—may be found with two times the above mentioned TWI. This situation can be interesting when speaking of children—house dust can be an important source of exposure—and smokers: in the last situation, the contribution of tobacco smoking can be similar if compared with dietary effects on the internal exposure. On these bases, the CONTAM Panel has concluded that the current exposure to Cd should be reduced [25].

Consequently, the EU Commission has set ML for Cd in a large variety of foodstuffs, in accordance with the Regulation (EC) No 1881/2006. At present, current ML have been revised by the Commission Regulation (EC) No 629/2008 [40].

Mercury

The importance of mercury (Hg) as an environmental contaminant is discussed for many years. It can be released by both natural and anthropogenic sources [85]. Natural emissions are substantially correlated with the degassing of the earth's crust and the activity of volcanoes. In addition, a certain amount of Hg may be emitted by simple water evaporation.

On the other hand, anthropogenic emissions—industrial activities, mining, etc.—appear more important. Moreover, Hg is continually mobilised in the atmosphere and superficial waters by means of repeated mobilisation and re-mobilisation cycles.

With reference to the most abundant form of this heavy metal in the environment, Hg is naturally emitted from land and ocean surfaces as elemental mercury. This form is the main part of the total quantity of existing Hg in the environment: industrial activities do not seem to emit remarkable amounts of non-elemental Hg.

However, the subsequent release of this metal on soils and waters should be discussed because Hg may be deposited as mercuric mercury (Hg^{2+}). With reference to aquatic systems, Hg can be found as elemental mercury and Hg^{2+}-complexes and organic mercury (methylmercury and dimethylmercury). The presence of these molecules is very important on the quantitative level: for example, methylmercury may be approximately 5 % of the total Hg content in marine waters. On the other hand, this percentage may arrive up to 30 % in fresh water [88].

Hg and methylmercury in particular have been extensively studied because of their health risk. With exclusive relation to the EU legislation, the CONTAM Panel has been asked to release in 2004 an Opinion on mercury and methylmercury in food [21]. However, the lack of scientific data has not allowed the reliable estimation of the intakes of high consumers in different populations, according to the CONTAM Panel [21].

At present, current ML for Hg are laid down in the Annex, Section 3, of the Commission Regulation (EC) No 1881/2006. This Annex has been recently amended by Commission Regulations (EC) No 629/2008 [40] and No 420/2011 [47].

It has to be considered that current ML values confirm the conclusion of the CONTAM Panel on mercury and methylmercury in food [21]. At present, ML values are established for mercury in fishery products and muscle meat of fish and in food supplements.

On the other side, performance characteristics for the analytical determination of mercury are set in the Regulation (EC) No 333/2007, recently amended by the Commission Regulation (EU) No 836/2011 [48].

On the other side, water is a different 'food product' because of the wide availability, the correlate meaning of 'publicly available good and service' and the easy contamination. With relation to 'drinking waters intended for human consumption', harmonised levels for Hg are set by the Council Directive 98/83/EC. In detail, ML values for Hg are 1 µg/l in 'water intended for human consumption'. The same ML has been imposed for mineral natural waters, in accordance with the Commission Directive 2003/40/EC. With reference to the analytical determination of mercury in waters, performance characteristics are set in the Council Directive 98/83/EC and in the Commission Directive 2003/40/EC.

A final mention should be made with relation to the presence of Hg in food additives: a ML between 0.1 and 3 mg/kg of Hg has been defined in accordance with the Commission Directive 2008/84/EC and subsequent amendments.

Tin

The importance of tin (Sn) as a food contaminant is known and discussed when speaking of canned foods because of the composition of 'tin cans' [80]. However, two reflections should be honestly made with reference to the so-called 'tin poisoning':

(a) Detected Sn levels in several canned foods can be dramatically increased depending on electrochemical properties of metallic cations (Pb) in ancient cans and nitrate in modern cans [82]
(b) Historically, the term 'tin poisoning' was correlated to Sn. On the contrary, gastric symptoms were probably correlated to Pb contents in old-style cans [82]. At present, Pb is no longer allowed in modern tin cans [80].

With exclusive reference to inorganic Sn, the SCF Panel has concluded in its opinion of 12th December 2001 that optimal ML for this element should be lower than 150 mg/kg in canned beverages and 250 mg/kg in other canned foods because of the possible occurrence of gastric irritation in some individuals [84]. As a result, ML levels for inorganic tin in these foods have been decided as follows, including also specific levels for foods intended to infants and young children on a 'precautionary basis'.

In detail, ML for tins are provided in the Commission Regulation No 1881/2006, Annex, Section 3 (metals), with reference to following products:

- Canned foods other than beverages
- Canned beverages, including fruit juices and vegetable juices
- Canned baby foods and processed cereal-based foods for infants and young children, excluding dried and powdered products
- Canned infant formulae and follow-on formulae (including infant milk and follow-on milk), excluding dried and powdered products
- Canned dietary foods for special medical purposes intended specifically for infants, excluding dried and powdered products.

With the exception of the first two food categories (ML are defined 200 and 100 mg/kg wet weight respectively), the 'basic' ML is generally 50 mg/kg.

Sampling methods and analytical requirements, including the accreditation for laboratories, have to be performed in accordance with the Regulation (EC) No 333/2007 [39] as recently amended by the Commission Regulation No 836/2011 [48].

2.5.2.4 Residual Pesticides

At present, different substances and mixtures may be defined 'pesticides'. On the regulatory viewpoint, following chemicals and chemical mixtures are classified pesticides in the EU at least:

- Insecticides, acaricides
- Herbicides, fungicides, plant growth regulators
- Rodenticides
- Biocides
- Veterinary medicines.

The above mentioned subdivision in classes depends on the peculiar use: crop protection, control of insects, promotion of the plant growth, food preservation [79]. In detail, pesticides are chemical substances with the ability of damaging vital functions of several life forms. For example:

(a) Insecticides and acaricides can be used with the aim of eradicating the presence of insects and mites respectively
(b) Fungicides can be used with the aim of eradicating the presence of fungi
(c) Herbicides eliminate the presence of certain plants
(d) Rodenticides are used against rodents.

The situation of so-called 'biocides' is different. The authorisation of these substances is carried out in accordance with the Directive 98/8/EC [54]. Basically, the main difference is the definition of biocides: these chemical or biological products are produced with the clear aim of destroying, eradicating or prevent the action of dangerous life forms [76]. Four product typologies have been established:

• Disinfectants and general biocidal products
• Preservatives
• Pest control products
• Other biocidal products.

Two reflections should be made. First of all, above mentioned categories are wide enough [76]:

(1) The category of disinfectants and general biocides include human hygiene products, disinfectant products for cleaning private and public health areas, veterinary hygiene products, disinfectants for food and feed areas, disinfectants for drinking water
(2) The group of preservatives is substantially destined to professional uses such as the protection of wood, fibres, masonry, etc.
(3) Pest control products (widely used in the food industry) include rodenticides, molluscicides, avicides, piscicides and a wide selection of products against insects, mites and other arthropods (insecticides, acaricides, repellents, attractants)
(4) Finally, the group of 'other biocidal products' include preservatives for food and feedstock (these products are not considered in the second category), products for the limitation of the presence of different vertebrates, etc.

As a clear consequence, pesticides and biocides are two different groups with some overlap: at the same time, Directives 91/414/EEC and 98/8/EC can be applied with reference to the same substance. The main difference is always the declared use of the product: plant protection products are used with the aim of limiting or eradicating organisms with dangerous effects on plants or plant products. On the other side, biocides are produced and used with the aim of limiting or eradicating dangerous organisms for the human health or non-vegetable products [76]. In addition, the Directive 98/8/EC might be applied with relation to several medicines, veterinary medicines and cosmetics. Actually, these products should fall [76] within the scope of Directives on medicines, veterinary medicines and cosmetics (Sect. 2.5.2.5).

With the exception of biocides, the use of pesticides is always correlated with the promotion of agriculture and horticulture, although a certain number of non-agricultural pesticides may be authorised [79]. As a consequence, two main questions may be made when speaking of pesticides [79]:

- The usage of these substances on a massive scale has sometimes caused poisoning incidents in human beings and domestic animals, with environmental adverse effects
- The authorisation for use of 'pesticides' is always correlated with the evidence of possible effects on humans, animals and the environment (water, soil, air, non-target organisms). In addition, effects on plants have to be evaluated and the efficacy of pesticides has to be assessed.

For these reasons, harmonised rules are needed with reference to the placing on the market of plant protection and non-agricultural products. Consequently, only authorised pesticides may be used in the EU because of the necessity of demonstrable evidences about above mentioned points.

Historically, the first document with regulatory importance has been the Council Directive 91/414/EEC [7] concerning the placing of plant protection products on the market. This document, entered into force in 1993, has given harmonised rules with concern to plant protection products and active substances. However, the basic system of the above mentioned Directive has been repealed after 13 years [76] by the Regulation (EC) No 1107/2009 of the European Parliament and of the Council of 21 October 2009 concerning the placing of plant protection products on the market [69].

In accordance with the existing legislation in the EU, residual pesticides correspond to traces of pesticides in treated products. In detail, the Directive 91/414/EEC states (article 2, point 1) that 'pesticide residues' means residues present in or on the surface of products covered by the Annex I to this document. Residuals pesticides can also include active substances, metabolites and/or breakdown or reaction products of active substances currently or formerly used in plant protection products. This definition concerns also all possible substances generated as the result of use in plant protection, in the veterinary medicine and as a biocide [69, 70].

In addition, the highest level of a residual pesticide in or on the surface of foods or feeds is defined 'maximum residue level' (MRL) on condition that pesticides are applied correctly. This definition concerns also the existence of 'good agricultural practices' (GAP). In detail, GAP are considered in relation to the definition of maximum application rates, the number of applications and minimum pre-harvest intervals for the satisfactory control of a defined pest in similar areas [79].

Basically, the EFSA must verify that each examined residual pesticide is safe for all EU consumer groups, including vulnerable consumers and categories such as babies, children and vegetarians. Should some type of risk be evaluated and established for one or more specific consumer groups, the request for a MRL would be rejected. As a consequence, the examined pesticide could not be used on the specified crop.

On these bases, the EU Commission can set a new MRL for the specific pesticide with the concomitant amendment or the removal of the existing value after the request of a scientific Opinion to the EFSA. This step has to be performed by means of a specific Regulation fixing MRL for foods and animal feeds.

On the other side, 'active substances' have to be assessed when contained in a specific pesticide or biocide. Firstly, the EU Commission has to approve these active substances; as a result, the active substance is included in the Annex I of the respective Directive. Subsequently, Member States can authorise the use of approved active substances on their territories on condition that these chemicals remain compliant with EU rules. The authorisation procedure is defined in accordance with the Commission Regulation (EU) No 188/2011 [49].

At present, MRL for pesticide residues are listed in the Regulation (EC) No 396/2005 [63] and subsequent amendments. With the exception of peculiar substances, the 'default' MRL is value of 0.01 mg/kg according to the Art 18(1) (b) of the Reg. No 396/2005. In addition, the updated list of pesticides and related MRL can be found in a dedicated 'EU Pesticides' database.[3] The number of approved active substances and pesticides is very notable: for this reason at least, there is not a real convenience in showing a long list of approved chemicals and mixtures. However, two examples can be shown here with reference to active substances and pesticides.

The above mentioned 'EU Pesticides database' is subdivided in two sections depending on the main scope—active substances or pesticides. When speaking of active substances, the database user can ask for information about a specified and listed substance such as 1,3-diphenyl urea. The database can show following data:

- Status of approval with reference to a specific document (1,3-diphenyl urea is not approved at the date of 24th June 2014)
- The related category (plant growth regulator)
- The classification (1,3-diphenyl urea is not classified at the date of 24th June 2014)
- Available toxicological information (absence of information at the date of 24th June 2014)
- European MRL. At present, the 'default' MRL of 0.01 mg/kg is applied.

When speaking of pesticides, the dedicated section of the database provides two options. The database user can ask for information about a specified and listed pesticide or about pesticides for a selected food product. Should the user ask for information about the product: cardamom, the database would show a list of approved pesticides with correlated MRL. In this situation, the list includes at least: 1,3-dichloropropene, 2-phenylphenol and barban (fat soluble) with MRL of 0.05, 0.1 and 0.1 mg/kg respectively.

---

[3] This database of the Directorate General for Health & Consumers (DGSANCO) is available at the link: http://ec.europa.eu/sanco_pesticides/public/?event=homepage Actually, two different databases are available at this web address: a list of active substances in accordance with the Regulation (EC) No 1107/2009 and the Pesticides database in accordance with the Regulation (EC) No 396/2005. However, it has to be remembered that above mentioned databases have no legal values. Official MRL are listed exclusively in the official legislation.

On the other side, the database user can ask for information about a specific pesticide. Should this substance be barban (fat soluble), the database would show the related MRL for every specific group and example of individual products to which the MRL can apply. In addition, the reference to Annexes (II and III B) of the Reg. No 396/2005 would be shown with an additional link to the last legislation of reference. With concern to barban, the Commission Regulation (EC) No 149/2008 amending Regulation (EC) No 396/2005 has to be considered.

Finally, legal rules on setting MRL, responsibilities for the private sector, the surveillance on the market by Member States are defined by sectoral norms.

2.5.2.5 Residues of Veterinary Medicines

A sub-group of chemicals can be detected in foods of animal origin when food-producing animals are treated with medicines. The aim of similar practices is the prevention or the treatment of animal diseases. However, it is well known that similar medicines can release residual chemicals in food products obtained from treated animals. For this reason, the risk assessment evaluation has to assure that the inevitable presence of these residuals does not harm the final consumer.

Actually, medicinal residues may be confused with pesticides (Sect. 2.5.2.4). In fact, several of these chemicals may be also used as veterinary medicines for animal treatment, including also the addition of pesticide substances in feedingstuffs. In these situations, the assessment of above mentioned veterinary residues may be difficult with reference to the determination of MRL (Sect. 2.5.2.4). In relation to these limits, the distribution and the metabolism of residues can depend also on the physical placing of chemicals in the body of animals—fatty tissues, muscles, etc. [75].

The matter of veterinary residues has been firstly carried out in the EU by means of the Council Directive 96/23/EC concerning rules on veterinary medicines, pesticides and contaminants in food of animal origin [9].

With reference to specific obligations, the previous legislation stated that:

1. Member States are obliged to implement residue monitoring plans for the unauthorised use of substances, the abuse of authorised veterinary medicines and the reduction of detectable residues
2. Extra EU-countries exporting to the EU are forced to guarantee an equivalent level of food safety when speaking of imported animal food to the EU by means of the implementation of a residue monitoring plan
3. In addition, Member States have to assure that FBO carrying on the initial processing of products from treated animal and correlated farms are in full compliance with existing laws in the EU.

The above mentioned Council Directive 96/23/EC gave also clear indications with relation to official controls, the establishment of national reference laboratories and duties of official EU veterinarians [9].

Subsequently, the Council Regulation (EEC) No 2377/90 gave the definition of MRL for the safety evaluation of residues of veterinary medicinal products.

Finally, the Regulation (EC) No 470/2009 is applied at present with reference to procedures for the establishment of residue limits of pharmacologically active substances in foodstuffs of animal origin [70].

The safety assessment of residues has to be performed by a specific Committee for Medicinal Products for Veterinary Use (CVMP). Should a MRL opinion be given by the CVMP, the European Medicines Agency would receive it without prejudice for the final Commission Regulation. After the publication of the Commission Regulation, the MRL can be also justified with the correlated 'European Public MRL Assessment Report'. For example, the CVMP had concluded in 2008 that a specific MRL is not needed [10] for a residue such as 1-methyl-2-pyrrolidone [10].

As a consequence, the mention of specific MRL values for every allowed pharmacological substance in foodstuffs of animal origin is provided by the Commission Regulation (EU) No 37/2010 [44]. With reference to this document, it has to be clarified that the list of allowed substances concerns both veterinary residues and biocidal products (Sect. 2.5.2.4). Once more, the difference is dependent on the declared use: veterinary medicines are applied directly to the animal while biocides may or may not be used in the same way. Anyway, both classes are pharmacologically active substances and their classification is correlated with MRL.

For example, the above mentioned 1-methyl-2-pyrrolidone is present in the Annex to this Regulation without MRL. On the other hand, amoxicillin—a well known anti-infectious agent and antibiotic—can be allowed on condition that MRL is 4 μg/kg for milk and 50 μg/kg for following animal parts: muscle, fat, liver and kidney [43, 44].

Finally, prescribed analytical methods have to comply with the Regulation (EC) No 882/2004. Only EU reference laboratories designated by the Commission may be consulted by the European Medicines Agency. Reg. (EC) No 882/2004 is a very general Regulation and has to be considered when speaking of official controls in a large variety of foodstuffs [61].

2.5.2.6 Other Contaminants. ML for Monochloropropane-1,2-Diol

With relation to 3-MCPD esters, ML of 0.02 mg/kg have been established for hydrolysed vegetable proteins and soy sauces at a community level in accordance with the last statement of the CONTAM Panel [24], although current discussions at the Codex Alimentarius propose higher maximum levels of 0.4 mg/kg for liquid condiments.

2.5.2.7 Other Contaminants. ML for Polycyclic Aromatic Hydrocarbons

PAH are known as food contaminants because of their presence during smoking, heating and drying processes. Combustion products may come into direct contact with foods. Moreover, fish and fishery products above all can be contaminated with PAH by environmental pollution.

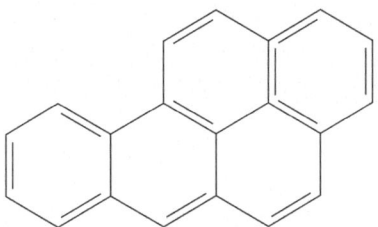

Fig. 2.2 The chemical structure of benzo[*a*]pyrene, molecular formula: $C_{20}H_{12}$, molecular weight: 252.31 g/mol, Chemical Abstracts Service number: 50-32-8. This molecule belongs to the group of polycyclic aromatic hydrocarbons (PAH). It is widely recognised as a potent carcinogen and mutagen agent in different food and non-food products, including tobacco cigarettes. For these reasons, benzo(a)pyrene is currently used as a marker for adverse effects of carcinogenic PAH in foods. BKchem version 0.13.0, 2009 (http://bkchem.zirael.org/index.html) has been used for drawing this structure

One of PAH, benzo[*a*]pyrene (Fig. 2.2), is currently used as a marker for adverse effects of carcinogenic PAH in foods. Actually, PAH include other chemicals: benzo[*a*]anthracene, benzo[*b*]fluoranthene, benzo[*j*]fluoranthene, benzo[*k*]fluoranthene, benzo[*g,h,i*]perylene, chrysene, etc. [4].

At present, ML have been defined for PAH—benzo[*a*]pyrene—in several foodstuffs, including meat and meat products, fish and fishery products, milk and milk products, oils and fats. Once more, the framework EU document is the Commission Regulation (EC) No 1881/2006 (ML for chemical contaminants in foodstuffs), Annex, Section 6. Other food groups may be added to the list of ML at a later stage [72] because of the lack of available data for cocoa butter and food supplements, etc.

In addition, a ML of 10 ppb for benzo[*a*]pyrene and 20 ppb for benzo[*a*] anthracene have been defined in smoke flavourings by the Regulation 2065/2003 (Sect. 2.4.1).

Actually, proposed official ML are very low if compared with current available data. The reason is the necessity of assuring that the health of consumers is not affected by consuming these products. For this reason, national authorities perform the routine surveillance of foods and laboratory analyses on samples of potentially contaminated produce with the aim of determining PAH levels in products [72].

Finally, sampling procedures for the official control of the levels of benzo[*a*] pyrene in foods have to be carried out in accordance with the Annex to the Regulation (EC) No 333/2007.

2.5.2.8 Other Contaminants. ML for Dioxins and Polychlorinated Biphenyls

With concern to dioxins and PCB, the last amendment to existing ML values is the Commission Regulation (EU) No 277/2012 [52]. This norm has amended the previous and applicable Directive 2002/32/EC [56].

Lists of ML in the Annexes I and II to Directive 2002/32/EC contain many substances. Dioxins correspond to a group of 75 polychlorinated dibenzo-*para*-dioxin (PCDD) congeners and 135 polychlorinated dibenzofuran (PCDF) congeners. At present, 17 of these substances are judged 'of toxicological concern'.

On the other hand, PCB are 209 different congeners. At present, this category can be subdivided in two groups depending on recognised toxicological properties. Moreover, 12 of these congeners are very similar to dioxins from the toxicological viewpoint: for this reason, there are 12 'dioxin-like' PCB.

ML for dioxins and PCB are defined in function of two basic factors:

(a) The classification of undesirable substances, and
(b) The peculiar group of feedingstuffs.

As a result, undesirable substances are listed as follows [52, 56]:

• 'Dioxins'. This term is for 'sum of PCDD and PCDF'
• Sum of 'dioxins' and 'dioxin-like PCB'. Last words are for: 'sum of PCDD, PCDF and PCB'
• Non-dioxin-like PCB.

Interestingly, ML values for these substances are expressed in two ways:

• With relation to the first two mentioned groups, ML are expressed in ng 'TEF'/ kg (ppt) relative to a feed with a moisture content of 12 %, where 'TEF' is for: 'toxic equivalent factor'
• With reference to non-dioxin-like PCB, ML values are set in μg/kg (ppb) relative to a feed with a moisture content of 12 %.

This situation means also that the first two groups of chemicals are expressed themselves as TEF [50–52]. This additional measure concerns all the individual dioxin and dioxin-like PCB congeners of toxicological concern.[4]

2.5.2.9 A Final Note About Official Controls for Contaminants, Residues of Veterinary Medicines and Pesticides

The Regulation (EC) No 882/2004 (official controls on foodstuffs in the EU) considers the necessity to maintain in force specific rules in the area of feed and food and animal health. Actually, official controls concern also residues of veterinary medicines and pesticides as regulated by the Council Directive 96/23/EC (substances with pharmacological action) and the Regulation (EC) No 396/2005 (ML for pesticides).

In detail, the Directive 96/23/EC declares that Member States have to implement national residue control plans [9]. These plans and related results (with actions taken as follow-up of non-complaint results) have to be submitted annually

---

[4] The interpretation of results is not possible without a dedicated table of TEF for dioxins, furans and dioxin-like PCB. This table is derived from the World Health Organization [89].

to the Commission. At the same time, the minimum number of samples is defined by the same document for residues of veterinary medicines, pesticides and environmental contaminants among the group of substances to be controlled in the framework of residue control plans.

As recently observed [42], the possibility of integrating currently applicable rules to official controls on pesticides, contaminants and residues of pharmacologically active substances in the more general framework of the Reg. (EC) No 882/2004 should be considered. Naturally, the basic aim should be the integration of both sectors (normal controls on foods and feeds/controls on contaminants) into Member States' multi-annual control plans.

2.6 The REACH Regulation

By the regulatory viewpoint, the management of chemicals in food and feed sectors has been profoundly modified because of a new legislation concerning the registration and the use of chemical substances in human activities.

With the exception of regulatory documents dealing with emerging food and feed-related risks, the so-called 'chemical risk'—one of the three basic pilasters of the 'Hazard Analysis and Critical Control Points' approach in the food sector—is regulated by the Regulation on 'Registration, Evaluation, Authorisation and Restriction of Chemicals', also named REACH [64].

In accordance with the REACH, the safety evaluation of chemical substances in all industrial fields cannot be assured without solid bases such as selected methods or procedures. The absence of similar documents has determined the decision of the EU Legislator in relation to the creation of a four steps-procedure for the evaluation of chemicals [81].

The REACH considers four steps with reference to the identification of chemicals and correlated concerns in the EU [64, 81]:

(a) The preventive registration
(b) The evaluation
(c) The final authorisation
(d) The determination of possible restrictions for specific chemical substances, depending on the declared use or uses.

Basically, the REACH is a horizontal legislation because of the general application to all chemical substances in the EU. As stated in the Article 1, the aim of the REACH is the promotion of high safety and protection levels for the human being and the environment. Moreover, the free circulation of (approved) chemical and non-chemical substances on the internal market has to be assured.

The REACH has to be applied when speaking of manufacture, placing on the market or use of 'substances', preparations or articles realised with these 'substances' [64].

With reference to 'substances', the REACH 'shall not apply to (Article 2):

- Radioactive substances within the scope of Council Directive 96/29/Euratom of 13 May 1996
- Substances, on their own, in a preparation or in an article, which are subject to customs supervision, provided that they do not undergo any treatment or processing, and which are in temporary storage, or in a free zone or free warehouse with a view to re-exportation, or in transit
- Non-isolated intermediates
- The carriage of dangerous substances and dangerous substances in dangerous preparations by rail, road, inland waterway, sea or air'.

In addition, waste cannot be defined 'substance' in accordance with REACH, Art. 2(2) when speaking of the field of application of the Directive 2006/12/EC. Member States may decide detailed exemptions for specific substances in the interests of defence in accordance with the REACH, Art. 2(3).

Other exemptions are forecasted when 'substances' are used in:

(a) Medicinal products for human or veterinary use
(b) Food or feedingstuffs in accordance with the Regulation (EC) No 178/2002.

The provisions of the Title IV of the REACH (information in the supply chain) are not considered for following preparations in the finished state, intended for the final user in accordance with Art. 2(6):

(1) Medicinal products for human or veterinary use
(2) Cosmetic products
(3) Medical devices 'which are invasive or used in direct physical contact with the human body'
(4) Foods or feedingstuffs in accordance with the Regulation (EC) No 178/2002.

Finally, several exemptions from the obligation to register are allowed when speaking of substances mentioned in Annexes IV and V of the REACH, and other specific situations in accordance with the Article 2(7). Two reflections can be made:

(a) The Annex IV mentions substances that are used or may be also used as food additives (examples: distilled water; sunflower oil; sodium stearates; etc.)
(b) The Annex V mentions mainly substances that can be obtained incidentally from chemical reactions because of different reasons. In other words, these 'substances' are not intentionally used or added. Interestingly, chemical causes for unexpected reactions can be determined by the presence of chemical additives for the industry such as stabilisers, colorants, flavouring agents, antioxidants, fillers, plasticisers, corrosion inhibitors, desiccants, emulsifiers, pH neutralisers, fire retardants, lubricants, quality control reagents, etc.

On these bases, it can be inferred that a few substances only should be mentioned in the food sector when in connection to the REACH. On the other side, the correct definition of 'substance' (Art. 3) concerns every 'chemical element and its compounds in the natural state or obtained by any manufacturing process, including any additive necessary to preserve its stability and any impurity deriving from the process used, but excluding any solvent which may be separated without affecting the stability of the substance or changing its composition'.

From the viewpoint of FBO, the matter of REACH can be extremely difficult. Basically, the main question may be: 'Could any specific additive, flavouring or substance intended for food and/or feed production or storage be considered in the field of application of the REACH?' Actually, the Article 2 excludes similar possibilities. Two options remain to be discussed:

- The nature of chemical compounds that can be found in foods or feeds because of migration from food packaging materials
- The nature of food additives or similar substances when used in a different way instead of the intended use by an actor in the supply chain, in accordance with the Art. 3 (26).

Should these options be discussed, the FBO would be considered as a downstream user (DU) with correlated rights and obligations in accordance with the Title V: request of information, preparation of exposure scenarios for the identified use(s) and for any use outside the conditions described in an exposure scenario (Annex XII), etc.

The detailed description of the REACH is not one of the basic aims of this book. However, it may be clarified here that the application of REACH should concern at least:

- Food packaging materials (FPM) because of the presence of chemical materials and intermediates
- Active packaging devices. The scope of these systems is the modification of one or more food properties by means of the release of peculiar chemical compounds or the insertion of active and antiseptic principles [81]
- FPM from recycled materials.

These questions are briefly mentioned in Sect. 2.8. Naturally, the position of food packaging producers as DU at least may be easily understood while the role of FBO can be defined 'unclear'.

With reference to other 'active' players of the REACH, the European Chemicals Agency (ECHA) has been established with the aim of carrying out and promoting the above mentioned four steps-procedures. One of the most important results of the REACH is surely the creation of two different lists.

The first and most important of these lists is surely the 'Authorisation List' (Annex XIV) containing several 'substances of very high concern' (SVHC). At present, this list is available at the following link: http://echa.europa.eu/addressing-chemicals-of-concern/authorisation/ recommendation-for-inclusion-in-the-authorisation-list/authorisation-list.

According to the REACH, these substances 'are recognised:

- Carcinogenic, mutagenic or toxic to reproduction
- Persistent, bioaccumulative and toxic or very persistent and very bioaccumulative

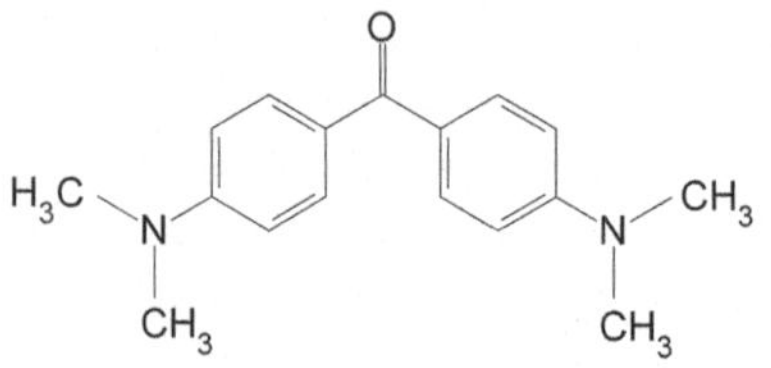

Fig. 2.3 The chemical structure of 4,4′-bis (*N,N*-dimethylamino) benzophenone, also named 'Michler's ketone', molecular formula: $-C_{17}H_{20}N_2O$, molecular weight: 268.4 g/mol, Chemical Abstracts Service number: 90-94-8. BKchem version 0.13.0, 2009 (http://bkchem.zirael.org/index.html) has been used for drawing this structure

- Identified, on a case-by-case basis, from scientific evidence as causing probable serious effects to human health or the environment of an equivalent level of concern as those above (e.g. endocrine disrupters)'.

The second list, also named 'Candidate List', mentions a broad selection of chemicals with different importance. Basically, all these substances are considered SVHC. The ECHA can submit detailed recommendations to the EU with the aim of including SVHC of the Candidate List in the Authorisation List.

For example, the 'Michler's ketone'—4,4'-bis (*N,N*-dimethylamino) benzophenone, Fig. 2.3—is currently a 'candidate' SVHC; the introduction of this chemical in the Authorisation list may be expected in the future. The same procedure has been carried out in the past for lead sulfochromate yellow, lead chromate and lead chromate molybdate sulphate red: all these substances are in the Authorisation list.

The importance of the 'Authorisation' status is easily comprehensible: the use of these substances may be prohibited in the EU or restricted when speaking of specific situations, uses and products. Anyway, the inclusion of a substance in the Candidate List is always a sort of 'alert' for all interested players, including also DU. The most recent addition to the Candidate List concerns [15]:

- 1,2-Benzenedicarboxylic acid, dihexyl ester, branched and linear
- Sodium perborate; perboric acid, sodium salt
- Sodium peroxometaborate
- Cadmium chloride.

2.7 Food Packaging Materials. Connections with Food Contamination

Basically, the regulatory profile of FPM is defined in the EU by the framework Regulation (EC) 1935/2004 on materials and articles intended to come into contact with food [62]. Different FPM can be regulated by a variety of specific documents: all these regulatory norms are listed in the framework Regulation.

At present, the situation in the EU can be summarised as follows:

- The sector of ceramic FPM is regulated by the Directive 84/500/EEC with exclusive reference to migration limits for Cd and Pb when released from decoration and/or glazing
- Films made of regenerated cellulose are regulated by the Directive 2007/42/EC (it contains the list of authorised substances and conditions for their use, including also provisions for plastic-coated regenerated cellulose films)
- Recycled plastics for the production of FPM are covered by the Regulation (EC) No 282/2008 (it contains requirements for the compliance of recycled plastics for the production of FPM and the authorisation procedure of recycling processes)
- The 'smart' packaging sector—active and intelligent materials and articles—is covered by the Regulation (EC) No 450/2009
- The wide sector of plastic materials has been recently covered by the Regulation (EU) No 10/2011 as amended by Reg. (EU) No 321/2011 and Reg. (EU) No 1282/2011.

Finally, several documents—the Directive 93/11/EEC and the Regulation 1895/2005/EC—are still valid with relation to the detection and related limits of nitrosamines and several epoxy derivatives respectively. This situation does not take into account the number of national laws with concern to uncovered FPM and articles in the EU ambit.

The problem of FPM in relation to foods can be seen by two different viewpoints.

First of all, can FPM release harmful substances? Should the answer be positive, the next step would be the definition of these substances and—where appropriate—the correct placement in the EU legislation. As above clarified (Sect. 2.6), 'harmful substances' may be seen in the ambit of REACH with consequent rights and duties for different players of the food and feed chain. However, the detection of 'food contaminants' such as Cd or Pb in foods packaged can be covered by the REACH with the possible exception of packaged products in ceramic containers. Should this be the situation, the Directive 84/500/EEC would apply. In spite of this argument, the REACH may be 'in force' when speaking of Cd or Pb migration (and exceeded limits) from ceramic objects. It can be concluded that the detection of 'harmful substances' exceeding defined migration limits (such as 'overall migration limit' and 'specific migration limits' for individual authorised substances by plastic materials) poses distinct problems on the regulatory level and by the safety viewpoints. The superposition of different regulatory documents may confuse FBO.

On the other hand, FPM can pose other hygiene and safety problems depending on the possibility of 'food packaging failures' and possible migrations of chemical substances. These words include different defects [80, 81]:

- Interactions between food and packaging with unexpected and possibly dangerous consequences
- Defects caused by improper FPM design (when the container is used for the 'wrong' food)

- Failures caused by FPM defects without food interactions
- Microbiological contaminations caused by improper storage and/or use of FPM near food producers or packers
- And other food-related failures.

Anyway, the occurrence of food-related failures with detection of undesired and/ or dangerous chemical intermediates by FPM has to be examined within a well-determined regulatory framework. As above explained, the detection of chemical intermediates may be seen in foods with relation to the (presumptive) origin:

- The presence of chemicals poses a food safety problem in the EU ambit and should be examined in accordance with the Reg. (EC) No 882/2004. This interpretation includes also the detection of food additives and different contaminants (Sect. 2.2–2.5)
- Alternatively, the detection of chemicals poses a general safety problem in the EU ambit and should be examined in accordance with the REACH (Sect. 2.6)
- The third interpretation considers the role of harmful substances outside of the scope of the REACH. Should this be the situation because of possible exceptions (Sect. 2.6), the safety problem would be considered in the EU ambit in accordance with specific legislations on FPM.

The detailed discussion of food-related failures caused by the interaction between foods and FPM is not the basic aim of this book. The interested Reader is invited to consult more specific literature and regulatory texts. However, it should be remembered that many specific problems may be discussed by two different viewpoints.

A simple and hypothetical example can be easily made with reference to the presence of food additives (or whitening agents) such as titanium dioxide in packaged dried (powdered) cheeses. Should this chemical substance be originated by FPM (three-piece metal cans may be coated with white epoxy-phenolic enamels: the pigment, titanium dioxide, is dispersed in these enamels), the discussion would take into account specific packaging-related provisions, where available, without connection to REACH: titanium dioxide should not be expected in foods by FPM migration.

On the other hand, titanium dioxide, also named E171, may be added in several foods as a specific additive [11]. Should this be the situation, the discussion would take into account specific food-related provisions and limits, where available, with reference to the Reg. (EC) No 882/2004 and without connection to REACH. E171 is 'limited by GMP' in cheeses according to the Codex Alimentarius Commission.

2.8 Chemical Hazards and Crisis Management. Lessons from Experiences

At present, the management of food-related crises and alarms in the EU can be complicated enough because of the number of presumptive emerging risks.

From a general and historical viewpoint, the following list of food-related topics might be taken into account:

(a) Bovine spongiform encephalopathy, also known as 'mad cow disease'
(b) Nanomaterials in foods
(c) Radioactive foods
(d) Unexpected presence of genetically modified organisms in foods and feeds
(e) Food adulterations
(f) Food contaminations by chemicals (examples: dioxins, etc.)
(g) Avian flu.

Actually, the number of presumptive food-related alarms in the EU and outside this economic region can be longer than this list. The main problem is the discrimination between presumptive alerts without solid scientific bases and real concerns.

With exclusive reference to the EU, the identification of 'emerging risks' should be clear enough. In accordance with the Regulation (EC) No 178/2002, art. 34, 'the Authority shall establish monitoring procedures for systematic searching for collecting, collating and analysing information and data with a view to the identification of emerging risks in the fields within its mission'.

As a consequence, the EFSA has recently established an 'Emerging Risks Exchange Network' (EREN) with the aim of exchanging information between EFSA and Member States on possible emerging risks for food and feed safety. As a consequence, the EREN can be a reliable observer with reference to the current situation of emerging issues, including chemical hazards.

As recently stated [33], most frequently evaluated issues in the EU appear to be chemical contaminants while microbiological hazards seem to be less important on the statistic level. In addition, other analysed issues have concerned labelling issues, presumptive risks from allergens, antimicrobial resistance, new industrial processes and the increased presence of foreign bodies in packaged foods. Finally, the EREN has also considered in 2011 possible concerns with reference to the presence of substances with pharmaceutical properties in food supplements. In this situation, the risk is correlated with adverse reactions in consumers [33].

With exclusive reference to chemical hazards, the following situations have been investigated between 2010 and 2013 [87]:

• Accumulation of pharmaceuticals and personal care products in crops
• Bisphenol A in food contact materials
• Alternatives to bisphenol A
• Potential chemical contamination of food from recycled paper
• Masked mycotoxins
• 3D-food printing.

In addition, other emerging risks in the EU have been identified 33]:

• 3-MCPD in soy sauce
• Acrylamid in fried food
• Benzene in soft drinks

- Bisphenol A in canned foods
- Dioxins
- Furans in coffee
- 2-isopropyl thioxantone in infant formula
- Melamine in infant milk powder
- Semicarbazid in food
- Sudan I food colour in chilli products
- Aflatoxin in red pepper.

Clearly, above expressed risks have been investigated with different results depending on the reliability of alerts, the seriousness of safety risks, the availability of scientific data and the answer of involved stakeholders [33]. Anyway, a possible strategy for the early and prompt management of food crises may be established on the basis of the risk acceptability. In other words, the risk may be defined 'acceptable' if following conditions are satisfied [87]:

(a) The voluntarity is demonstrated

(b) There is no alternative
(c) The risk may be defined 'natural'
(d) Consequences are immediate (without delayed effects on future generations)
(e) The risk is considered 'common' or 'old' hazard
(f) The exposure is discontinuous
(g) Clear benefits are demonstrable when speaking of the presence of a certain chemical in foods or feeds
(h) The risk can be defined 'non-vital', reversible and 'for adults' only.

On the other hand, the risk may be defined 'unacceptable' if one or more of the following conditions are demonstrable [87]:

(1) The voluntarity is not demonstrated
(2) Alternatives are available at present
(3) The risk is 'created' (anthropogenic sources)
(4) Consequences are not immediate (delayed effects on future generations)
(5) The risk is considered 'dread' or 'new' hazard
(6) The exposure is continuous
(7) Clear benefits are not demonstrable when speaking of the presence of a certain chemical in foods or feeds
(8) The risk can be defined 'vital', irreversible and 'for children or the next generation'.

Should a chemical hazard be considered as a real and unacceptable risk, existing procedures would be performed with an acceptable management of safety consequences. However, several situations may involve a serious direct or indirect risk to the human health. Moreover, the risk may be perceived or publicised as such. Finally, there are situations which are not likely to be prevented, eliminated or reduced to an acceptable level by provisions in place [87].

As a consequence, a close cooperation between interested stakeholders—national and international authorities, laboratories, industry, science, consumers, and media—is needed. In addition, a real 'crisis communication strategy' has to be forecasted and put in place [87].

For example, 'old' contamination episodes can be easily correlated to methyl-mercury poisoning and the correlated detection of Hg in fish and fish products. This situation, often remembered as the 'Minamata disease' [14], is surely 'unacceptable with reference to the above mentioned strategy. On the other hand, a precise and correct management strategy was difficult in 1975.

On the opposite hand, a more recent example concerns the detection of non-authorized antibiotics such as chloramphenicol and nitrofurans in aquaculture, in imported shrimps and poultry products. This situation has been managed by the EU with a 'zero tolerance' strategy [78]:

(1) Temporary controls for the presence of antibiotics on imported shrimps from eastern countries such as China, Vietnam, Thailand, and Indonesia during 2001 and 2002
(2) Suspension of imports from China
(3) Revocation of above mentioned preventive measures during the period 2002–2004.

Various reasons have been considered for this strategy, including [78]:

• The absence of internationally harmonised ADI or MRL for nitrofuran antibiotics and chloramphenicol
• The incomplete information on toxic effects.

A similar situation cannot be defined 'natural'. In addition, the correlated risk may be defined 'continuous', derived from anthropogenic sources, with uncertain effects on future consumers, etc. The above mentioned 'incident' may be useful for future investigations and the establishment of a pro-active risk management. Main pilasters should be [78]:

• Efficient measures with relation to education and quality control
• The implementation of more stringent regulations, based on risk assessment and previous 'incidents'
• The pro-active identification of potential future hazards.

In general, these advices should be useful for every food or feed-related crisis or emerging risk of chemical nature at least.

References

1. Alder L, Korth W, Patey AL, Schee HA, Schoeneweiss S (2001) Estimation of measurement uncertainty in pesticide residue analysis. J AOAC Int 84(5):1569–1578
2. Barcelou DG (2008) Medical toxicology of natural substances: foods, fungi, medicinal herbs, plants, and venomous animals. Wiley, Hoboken

3. Beretta B, Gaiaschi A, Galli CL, Restani P (2000) Patulin in apple-based foods: occurrence and safety evaluation. Food Addit Contam 17(5):399–406. doi:10.1080/026520300404815

4. Bláha L, Kapplová P, Vondrácek J, Upham B, Machala M (2001) Inhibition of gap-junctional intercellular communication by environmentally occurring polycyclic aromatic hydrocarbons. Toxicol Sci 65(1):43–51

5. CEC (1984) Council Directive 84/500/EEC of 15 October 1984 on the approximation of the laws of the Member States relating to ceramic articles intended to come into contact with foodstuffs. Off J L277:12–16

6. Commission Staff Working Document (2007) Annex to the proposal for a Regulation of the European Parliament and of the Council on flavourings and certain food ingredients with flavouring properties for use in and on foods amending Council Regulation (EEC) No 1576/89, Council Regulation (EEC) No 1601/91, Regulation (EC) No 2232/96 and Directive 2000/13/EC. Impact Assessment, Brussels, 28.7.2006, COM(2006) 427 final, SEC(2006)1043. Available http://register.consilium.europa.eu/doc/srv?l=EN&t=PDF&f=S T+16176+2007+INIT. Accessed 20 Jun 2014

7. CEC (1991) Council Directive 91/414/EEC of 15 July 1991 concerning the placing of plant protection products on the market. Off J Eur Comm L230:1–32

8. CEC (1993) Council Regulation (EEC) No 315/93 of 8 February 1993 laying down community procedures for contaminants in food. OJ Eur Comm L 37:1–5

9. CEC (1996) Council Directive 96/23/EC of 29 April 1996 on measures to monitor certain substances and residues thereof in live animals and animal products and repealing Directives 85/358/EEC and 86/469/EEC and Decisions 89/187/EEC and 91/664/EEC. Off J L125:10–32

10. CVMP (2008) 1-methyl-2-pyrrolidone Summary Report. Committee for Veterinary Medicinal Products. Available http://www.ema.europa.eu/docs/en_GB/document_library/ Maximum_Residue_Limits_-_Report/2009/11/WC500015085.pdf. Accessed 24 Jun 2014

11. Codex Alimentarius Commission (1978) Codex Stan 283-1978: General Standard for Cheese, rev.1-1999, amd.3-2008. Codex Alimentarius—International Food Standards, Rome. Available http://std.gdciq.gov.cn/gssw/JiShuFaGui/CAC/CXS_A06e.pdf. Accessed 25 Jun 2014

12. Codex Alimentarius Commission (1995) Codex General Standard for Food Additives 192, last revision 2013. Joint FAO/WHO Food Standards Programme, FAO, Rome. Available http://www.codexalimentarius.net/input/download/standards/4/CXS_192e.pdf. Accessed 13 Jun 2014

13. DGSANCO (2014) Practical guidance for applicants on the submission of applications on food additives, food enzymes and food flavourings, Version 6, updated on 10 April 2014. Available http://ec.europa.eu/food/food/fAEF/docs/practical_guidance_en.doc. Accessed 11 Jun 2014-06-11

14. D'itra F M (1991). Mercury contamination—what we have learned since Minamata. In: Proceedings of the fourth symposium on our environment, Singapore, 21–23 May 1990, pp 165–182. Springer, Netherlands. doi:10.1007/978-94-011-2664-9_18

15. ECHA (2014) Inclusion of Substances of Very High Concern in the Candidate List for eventual inclusion in Annex XIV. European Chemicals Agency, Doc. ED/49/2014:1'4. Available http://echa.europa.eu/documents/10162/9e20e208-4e32-4d59-be50-b398ef2e07bf. Accessed 25 Jun 2014

16. EFSA (2013) EFSA completes full risk assessment on aspartame and concludes it is safe at current levels of exposure. Available http://www.efsa.europa.eu/en/press/news/131210.htm. Accessed 13 Jun 2014

17. EFSA (2014) Evaluation of the increase of risk for public health related to a possible temporary derogation from the maximum level of deoxynivalenol, zearalenone and fumonisins for maize and maize products. EFSA J 12(5):3699, 61 p. doi:10.2903/j.efsa.2014.3699. Available http://www.efsa.europa.eu/it/efsajournal/doc/3699.pdf. Accessed 20 Jun 2014

18. EFSA ANS (2009) Scientific opinion on the re-evaluation of quinoline yellow (E 104) as a food additive. EFSA J 7(11):1329, 40 p. doi:10.2903/j.efsa.2009.1329

19. EFSA ANS (2010) Scientific opinion on the use of basic methacrylate copolymer as a food additive on request from the European commission. EFSA J 8(2):1513, 23 p. doi:10.2903/j.efsa.2010.1513

20. EFSA ANS (2013) Scientific opinion on the re-evaluation of aspartame (E 951) as a food additive. EFSA J 11(12):3496, 263 p. doi:10.2903/j.efsa.2013.3496

21. EFSA CONTAM (2004) Scientific opinion on the risk for public health related to the presence of mercury and methylmercury in food. EFSA J 10(12):2985, 241 p. doi:10.2903/j.efsa.2012.2985. Available http://www.efsa.europa.eu/it/efsajournal/doc/2985.pdf. Accessed 20 Jun 2014

22. EFSA CONTAM (2006) Opinion of the scientific panel on contaminants in the food chain on a request from the commission related to ochratoxin a in food, adopted on 4 April 2006. EFSA J 365:1–56. Available http://www.efsa.europa.eu/en/efsajournal/pub/365.htm. Accessed 19 Jun 2014

23. EFSA CONTAM (2008a) Opinion of the scientific panel on contaminants in the food chain on a request from the European Commission to perform a scientific risk assessment on nitrate in vegetables. EFSA J 689: 1–79

24. EFSA CONTAM (2008b) Statement of the Scientific Panel on Contaminants in the Food chain (CONTAM) on a request from the European commission related to 3-MCPD esters. doi:10.2903/j.efsa.2008.1048. Available http://www.efsa.europa.eu/it/efsajournal/doc/1048.pdf. Accessed 24 Jun 2014

25. EFSA CONTAM (2009) Scientific opinion of the panel on contaminants in the food chain on a request from the European commission on cadmium in food. EFSA J 980:1–139. Available http://www.efsa.europa.eu/it/efsajournal/doc/980.pdf. Accessed 20 Jun 2014

26. EFSA CONTAM (2010a) Scientific opinion on possible health risks for infants and young children from the presence of nitrates in leafy vegetables. EFSA J 8(12):1935, 42 p. doi:10.2903/j.efsa.2010.1935. Available http://www.efsa.europa.eu/de/search/doc/1935.pdf. Accessed 27 Jun 2014

27. EFSA CONTAM (2010b) EFSA panel on contaminants in the food chain; statement on recent scientific information on the toxicity of ochratoxin A. EFSA J 8(6):1626, 7 p. doi:10.2903/j.efsa.2010.1626. Available online: www.efsa.europa.eu. Accessed 19 Jun 2014

28. EFSA CONTAM (2010c) Scientific opinion on lead in food. EFSA J 8(4):1570, 151 p. doi:10.2903/j.efsa.2010.1570. Available http://www.efsa.europa.eu/en/search/doc/1570.pdf. Accessed 20 Jun 2014

29. EFSA CONTAM (2011a) Scientific opinion on the risks for public health related to the presence of zearalenone in food. EFSA J 9(6):2197, 124 p. doi:10.2903/j.efsa.2011.2197. Available http://www.efsa.europa.eu/it/efsajournal/doc/2197.pdf. Accessed 20 Jun 2014

30. EFSA CONTAM (2011b) Scientific opinion on the risks for animal and public health related to the presence of T-2 and HT-2 toxin in food and feed. EFSA J 9(12):2481, 187 p. doi:10.2903/j.efsa.2011.2481. Available www.efsa.europa.eu/efsajournal. Accessed 20 Jun 2014

31. EFSA SCF (2004) Opinion of the Scientific Panel on Contaminants in the Food Chain on a request from the Commission related to Deoxynivalenol (DON) as undesirable substance in animal feed. EFSA J 73:1–42

32. Ehrlich V, Darroudi F, Uhl M, Steinkellner H, Zsivkovits M, Knasmueller S (2002) Fumonisin B1 is genotoxic in human derived hepatoma (HepG2) cells. Mutagenesis 17(3):257–260. doi:10.1093/mutage/17.3.257

33. EREN (2011) Technical report of EFSA. Annual report on the emerging risks exchange network 2011. Supporting publications 2012:EN-280. European Food Safety Authority, Parma. Available http://www.efsa.europa.eu/en/search/doc/280e.pdf. Accessed 25 Jun 2014

34. European Commission (2003) Commission Regulation (EC) No 1425/2003 of 11 August 2003 amending Regulation (EC) No 466/2001 as regards patulin. Off J Eur Union L203:1–3

35. European Commission (2006) Commission Regulation (EC) No 1881/2006 of 19 December 2006 as regards maximum levels for nitrates in foodstuffs. Off J Eur Union L364:5–24

36. European Commission (2006) Commission Regulation (EC) No 1883/2006 of 19 December 2006 laying down methods of sampling and analysis for the official control of levels of dioxins and dioxin-like PCBs in certain foodstuffs. Off J Eur Union L364:32–43

37. European Commission (2006) Commission Regulation (EC) No 1882/2006 of 19 December 2006 laying down methods of sampling and analysis for the official control of the levels of nitrates in certain foodstuffs. Off J Eur Union L364:25–31

38. European Commission (2006) Commission Regulation (EC) No 401/2006 of 23 February 2006 laying down the methods of sampling and analysis for the official control of the levels of mycotoxins in foodstuffs. Off J Eur Union L70:12–34

39. European Commission (2007) Commission Regulation (EC) No 333/2007 of 28 March 2007 laying down the methods of sampling and analysis for the official control of the levels of lead, cadmium, mercury, inorganic tin, 3-MCPD and benzo(a)pyrene in foodstuffs. Off J Eur Union L88:29–38

40. European Commission (2008) Commission Regulation (EC) No 629/2008 of 2 July 2008 amending Regulation (EC) No 1881/2006 setting maximum levels for certain contaminants in foodstuffs. Off J Eur Union L173:6–9

41. European Commission (2009) Commission Regulation (EC) No 1152/2009 of 27 November 2009 imposing special conditions governing the import of certain foodstuffs from certain third countries due to contamination risk by aflatoxins and repealing decision 2006/504/EC. Off J Eur Union L313:40–49

42. European Commission (2009b) Report from the Commission to the European Parliament and to the Council on the application of Regulation (EC) No 882/2004 of the European Parliament and of the Council of 29 April 2004 on official controls performed to ensure the verification of compliance with feed and food law, animal health and welfare rules. Commission of the European Communities, Brussels, 8 July 2009, COM(2009) 334 final. Available http://register.consilium.europa.eu/doc/srv?l=EN&f=ST%2011948%202009%20INIT. Accessed 24 Jun 2014

43. European Commission (2010) Commission Regulation (EU) No 105/2010 of 5 February 2010 amending Regulation (EC) No 1881/2006 setting maximum levels for certain contaminants in foodstuffs as regards ochratoxin A. Off J Eur Union L35:7–8

44. European Commission (2010) Commission Regulation (EU) No 37/2010 of 22 December 2009 on pharmacologically active substances and their classification regarding maximum residue limits in foodstuffs of animal origin. Off J Eur Union L15:1–72

45. European Commission (2011a) Questions and answers on food additives commission Européenne—MEMO/11/783 14/11/2011. Available http://europa.eu/rapid/press-release_MEMO-11-783_en.htm?locale=FR. Accessed 13 Jun 2014

46. European Commission (2011) Commission Regulation (EU) No 1258/2011 of 2 December 2011, amending Regulation (EC) No 1881/2006 as regards maximum levels for nitrates in foodstuffs. Off J Eur Union L320:15–17

47. European Commission (2011) Commission Regulation (EU) No 420/2011 of 29 April 2011 amending Regulation (EC) No 1881/2006 setting maximum levels for certain contaminants in foodstuffs. Off J Eur Union L111:3–6

48. European Commission (2011) Commission Regulation (EU) No 836/2011 of 19 August 2011 amending Regulation (EC) No 333/2007 laying down the methods of sampling and analysis for the official control of the levels of lead, cadmium, mercury, inorganic tin, 3-MCPD and benzo(a)pyrene in foodstuffs. Off J Eur Union 215:9–16

49. European Commission (2011) Commission Regulation (EU) No 188/2011 of 25 February 2011 laying down detailed rules for the implementation of Council Directive 91/414/EEC as regards the procedure for the assessment of active substances which were not on the market 2 years after the date of notification of that Directive. Off J Eur Union L53:51–55

50. European Commission (2012) Commission Regulation (EU) No 232/2012 of 16 March 2012 amending Annex II to Regulation (EC) No 1333/2008 of the European Parliament and of the Council as regards the conditions of use and the use levels for Quinoline Yellow (E 104), Sunset Yellow FCF/Orange Yellow S (E 110) and Ponceau 4R, Cochineal Red A (E 124). Off J Eur Union L78:1–12

51. European Commission (2012) Commission Regulation (EU) No 594/2012 of 5 July 2012 amending Regulation (EC) 1881/2006 as regards the maximum levels of the contaminants ochratoxin A, non dioxin-like PCBs and melamine in foodstuffs. Off J Eur Union L176:43–45

52. European Commission (2012) Commission Regulation (EU) No 277/2012 of 28 March 2012 amending Annexes I and II to Directive 2002/32/EC of the European Parliament and of the Council as regards maximum levels and action thresholds for dioxins and polychlorinated biphenyls. Off J Eur Union L91:1–7

53. European Commission (2013) Commission Implementing Regulation (EU) No 1321/2013 of 10 December 2013 establishing the Union list of authorised smoke flavouring primary products for use as such in or on foods and/or for the production of derived smoke flavourings. Off J Eur Union L333:54–67

54. European Parliament and Council (1998) Directive 98/8/EC of 16 February 1998 concerning the placing of biocidal products on the market. Off J Eur Comm L123:1–63

55. European Parliament and Council (2002) Regulation (EC) No 178/2002 of the European Parliament and of the Council of 28 January 2002 laying down the general principles and requirements of food law, establishing the European Food Safety Authority and laying down procedures in matters of food safety. Off J Eur Comm L31:1–24

56. European Parliament and Council (2002) Directive 2002/32/EC of the European Parliament and Council and of the Council of 7 May 2002 on undesirable substances in animal feed. Off J Eur Union L 140:10–22

57. European Parliament and Council (2003) Regulation (EC) No 2065/2003 of the European Parliament and of the Council of 10 November 2003 on smoke flavourings used or intended for use in or on foods. Off J Eur Union L309:1–8

58. European Parliament and Council (2003) Regulation (EC) No 1829/2003 of the European Parliament and of the Council of 22 September 2003 on genetically modified food and feed. Off J Eur Union L268:1–23

59. European Parliament and Council (2003) Regulation (EC) No 1830/2003 of the European Parliament and of the Council of 22 September 2003 concerning the traceability and labelling of genetically modified organisms and the traceability of food and feed products produced from genetically modified organisms and amending Directive 2001/18/EC. Off J Eur Union L268:24–28

60. European Parliament and Council (2004) Directive 2004/10/EC of the European Parliament and of the Council of 11 February 2004 on the harmonisation of laws, regulations and administrative provisions relating to the application of the principles of good laboratory practice and the verification of their applications for tests on chemical substances (codified version). Off J Eur Union L50:44–59

61. European Parliament and Council (2004) Regulation (EC) No 882/2004 of the European Parliament and of the Council of 29 April 2004 on official controls performed to ensure the verification of compliance with feed and food law, animal health and welfare rules. Off J Eur Union L165:1–141

62. European Parliament and Council (2004) Regulation (EC) No 1935/2004 of the European Parliament and of the Council of 27 October 2004 on materials and articles intended to come into contact with food and repealing Directives 80/590/EEC and 89/109/EEC. Off J Eur Union L 338:4–17

63. European Parliament and Council (2005) Regulation (EC) No 396/2005 of the European Parliament and of the Council of 23 February 2005 on maximum residue levels of pesticides in or on food and feed of plant and animal origin and amending Council Directive 91/414/EEC. Off J Eur Union L70:1–1766. Available http://ec.europa.eu/food/plant/pesticides/legislation/max_residue_levels_en.htm. Accessed 23 Jun 2014

64. European Parlament and Council (2006) Regulation (EC) No 1907/2006 of the European Parliament and of the Council of 18 December 2006 concerning the Registration, Evaluation, Authorisation and Restriction of Chemicals (REACH), establishing a European Chemicals Agency, amending Directive 1999/45/EC and repealing Council Regulation (EEC) No 793/93 and Commission Regulation (EC) No 1488/94 as well as Council Directive

76/769/EEC and Commission Directives 91/155/EEC, 93/67/EEC, 93/105/EC and 2000/21/EC on materials and articles intended to come into contact with food and repealing Directives 80/590/EEC and 89/109/EEC. Off J Eur Union L396:1–849

65. European Parliament and Council (2008) Regulation (EC) No 1331/2008 of the European Parliament and of the Council of 16 December 2008 establishing a common authorisation procedure for food additives, food enzymes and food flavourings. Off J Eur Union L354:1–6

66. European Parliament and Council (2008) Regulation (EC) No 1332/2008 of the European Parliament and of the Council of 16 December 2008 on food enzymes and amending Council Directive 83/417/EEC, Council Regulation (EC) No 1493/1999, Directive 2000/13/EC, Council Directive 2001/112/EC and Regulation (EC) No 258/97. Off J Eur Union L354:7–15

67. European Parliament and Council (2008) Regulation (EC) No 1333/2008 of the European Parliament and of the Council of 16 December 2008 on food additives. Off J Eur Union L354:16–53

68. European Parliament and Council (2008) Regulation (EC) No 1334/2008 of the European Parliament and of the Council of 16 December 2008 on flavourings and certain food ingredients with flavouring properties for use in and on foods and amending Council Regulation (EEC) No 1601/91, Regulations (EC) No 2232/96 and (EC) No 110/2008 and Directive 2000/13/EC. Off J Eur Union L354:34–50

69. European Parliament and Council (2009) Regulation (EC) No 1107/2009 of the European Parliament and of the Council of 21 October 2009 concerning the placing of plant protection products on the market. Off J Eur Union L309:1–50

70. European Parliament and Council (2009) Regulation (EC) No 470/2009 of the European Parliament and Council and of the Council of 6 May 2009 laying down Community procedures for the establishment of residue limits of pharmacologically active substances in foodstuffs of animal origin, repealing Council Regulation (EEC) No 2377/90 and amending Directive 2001/82/EC of the European Parliament and of the Council and Regulation (EC) No 726/2004 of the European Parliament and of the Council. Off J Eur Union L152:11–22

71. European Parliament and Council (2013) Regulation (EC) No 609/2013 of the European Parliament and of the Council of 12 June 2013 on food intended for infants and young children, food for special medical purposes, and total diet replacement for weight control and repealing Council Directive 92/52/EEC, Commission Directives 96/8/EC, 1999/21/EC, 2006/125/EC and 2006/141/EC, Directive 2009/39/EC of the European Parliament and of the Council and Commission Regulations (EC) No 41/2009 and (EC) No 953/2009. Off J Eur Union L181:35–56

72. Food Safety Authority of Ireland (2009) Polycyclic Aromatic Hydrocarbons (PAHs) in Food. Toxicology Factsheet Series issue no 1 May 2009. Available http://www.google.com/url?q= http://www.fsai.ie/workarea/downloadasset.aspx%3Fid%3D8416&sa=U&ei=Q7apU8qNN9 Cz0QWl4YDoBQ&ved=0CCEQFjAC&sig2=4fpH7jS2HPc1H3rYxn_2UQ&usg=AFQjCN GSO5Kt7tkH2ZXXzNhkgMkcM7rFdw. Accessed 24 Jun 2014

73. Food Safety Authority of Ireland (2014) Smoke flavourings. Available http://www.fsai.ie/legislation/food_legislation/flavourings/smoke_flavourings.html. Accessed 16 Jun 2014

74. Galloway JN, Leach AM, Bleeker A, Erisman JW (2013) A chronology of human understanding of the nitrogen cycle. Phil Trans R Soc B 368(1621):20130120. doi:10.1098/r stb.2013.0120

75. Harris CA, Hill ARC (2004) Variability of residues in unprocessed food items and its impact on consumer risk assessment. In: Marrs TC, Ballantyne B (eds) Pesticide toxicology and international regulation, vol 1. Wiley, Chichester

76. Hussey DJ, Bell GM (2004) Regulation of pesticides and biocides in the European Union. In: Marrs TC, Ballantyne B (eds) Pesticide toxicology and international regulation, vol 1. Wiley, Chichester

77. Jelinek CF, Pohland AE, Wood GE (1988) Worldwide occurrence of mycotoxins in foods and feeds–an update. J Assoc Off Anal Chem. 72(2):223–230

78. Kleter GA, Groot MJ, Poelman M, Kok EJ, Marvin HJP (2009) Timely awareness and prevention of emerging chemical and biochemical risks in foods: proposal for a

strategy based on experience with recent cases. Food Chem Toxicol 47(5):992–1008. doi:10.1016/j.fct.2008.08.021

79. Marrs TT, Ballantyne B (2004). Pesticides: an overview of fundamentals. In: Marrs TT, Ballantyne B (eds) Pesticide toxicology and international regulation, vol 1. Wiley, Chichester

80. Parisi S (2012) Food packaging and food alterations: the user-oriented approach. Smithers Rapra Technology, Shawbury

81. Parisi S (2013) Food industry and packaging materials—performance-oriented guidelines for users. Smithers Rapra Technologies, Shawbury

82. Reilly C (2002) Metal contamination of food: its significance for food quality and human health. Blackwell Science Ltd, Oxford

83. SCF (1996) Reports of the Scientific Committee for Food, 35th series, Opinion on aflatoxin, ochratoxin A and patulin, expressed on 23rd September 1994, p. 45. The European Commission. Available http://ec.europa.eu/food/fs/sc/scf/reports/scf_reports_35.pdf. Accessed 19 Jun 2014

84. SCF (2002) Opinion of the Scientific Committee on Food on acute risks posed by tin in canned foods (adopted on 12 December 2001). Available http://ec.europa.eu/food/fs/sc/scf/out110_en.pdf. Accessed 23 Jun 2014

85. Schuster PF, Krabbenhoft DP, Naftz DL, Cecil LD, Olson ML, Dewild JF, Susong DD, Green JR, Abbott ML (2002) Atmospheric mercury deposition during the last 270 years: a glacial ice core record of natural and anthropogenic sources. Environ Sci Technol 36:2303–2310. doi:10.1021/es0157503

86. Simon R, de la Calle B, Palme S, Meier D, Anklam E (2005) Composition and analysis of liquid smoke flavouring primary products. J Sep Sci 28(9–10):871–882. doi:10.1002/jssc.200500009

87. Szabó MS (2014) Emerging risks in the European Union. Paper presented at the IAFP European Symposium on Food Safety, Budapest, 07 May 2014. Available http://www.foodprotection.org/downloads/meetings/program-activities/programs/m-ria-szeitzn-szab-directorate-for-food-safety-risk-assessment-hungary-em-emerging-food-safety-risks.pdf. Accessed 25 Jun 2014

88. Ullrich SM, Tanton TW, Abdrashitova SA (2001) Mercury in the aquatic environment: a review of factors affecting methylation. Crit Rev Environ Sci Technol 31:241–293. doi:10.1080/20016491089226

89. Van den Berg M, Birnbaum LS, Denison M, De Vito M, Farland W, Feeley M, Fiedler H, Hakansson H, Hanberg A, Haws L, Rose M, Safe S, Schrenk D, Tohyama C, Tritscher A, Tuomisto J, Tysklind M, Walker N, Peterson RE (2006) The 2005 World Health Organization reevaluation of human and mammalian toxic equivalency factors for dioxins and dioxin-like compounds. Toxicol Sci 93(2):223–241. doi:10.1093/toxsci/kfl055